AF806543

Machine Translation: Technologies and Applications

Volume 7

Editor-in-Chief

Andy Way, ADAPT Centre, Dublin City University, Dublin, Ireland

Editorial Board

Sivaji Bandyopadhyay, National Institute of Technology Silchar, Silchar, India

This book series tackles prominent issues in MT at a depth which will allow these books to reflect the current state-of-the-art, while simultaneously standing the test of time. Each book topic will be introduced so it can be understood by a wide audience, yet will also include the cutting-edge research and development being conducted today.

Machine Translation (MT) is being deployed for a range of use-cases by millions of people on a daily basis. Google Translate and Facebook provide billions of translations daily across many languages. Almost 1 billion users see these translations each month. With MT being embedded in platforms like this, available to anyone with an internet connection, one no longer has to explain what MT is on a general level. However, the number of people who really understand its inner workings is much smaller.

The series includes investigations of different MT paradigms (Rule-Based MT, Statistical MT, Neural MT, and newer AI models of translation), quality issues (MT evaluation, Quality Estimation), modalities (Sign Language MT), MT in practice (Post-Editing and Controlled Language in MT), and legal and ethical issues related to MT, topics which cut across the full range of real-world MT workflows.

Many of the most significant recent advances in AI, such as sequence-to-sequence models and the Transformer architecture, originated in MT research, so MT has been hugely influential in the broader field of AI. Nowadays, MT is increasingly embedded in broader multimodal systems (like ChatGPT), where text, speech, image, and video are generated and interpreted together. This book series has spanned seven years, during which MT system design has changed considerably. While standalone NMT systems are still built, MT in current AI frameworks often no longer depends on parallel source-target data, reflecting a major shift in how translation systems are developed.

This broader context also explains why a book on *efficient large language models* appears in this series as its final volume. Though not strictly MT, LLM efficiency directly affects MT deployment and usability, especially as MT becomes part of larger, general-purpose AI systems, so MT practitioners as well as AI generalists will benefit considerably from this volume.

For inquiries, please contact the Series Editor,
Andy Way: andy.way@adaptcentre.ie

Peyman Passban · Andy Way ·
Mehdi Rezagholizadeh
Editors

Enhancing LLM Performance

Efficacy, Fine-Tuning, and Inference Techniques

Springer

Editors
Peyman Passban
Dublin City University
ADAPT Centre
Dublin, Ireland

Andy Way
Dublin City University
ADAPT Centre
Dublin, Ireland

Mehdi Rezagholizadeh
Huawei Technologies
Noah's Ark Lab
Concord, ON, Canada

ISSN 2522-8021 ISSN 2522-803X (electronic)
Machine Translation: Technologies and Applications
ISBN 978-3-031-85746-1 ISBN 978-3-031-85747-8 (eBook)
https://doi.org/10.1007/978-3-031-85747-8

© The Editor(s) (if applicable) and The Author(s), under exclusive license to Springer Nature Switzerland AG 2025

This work is subject to copyright. All rights are solely and exclusively licensed by the Publisher, whether the whole or part of the material is concerned, specifically the rights of translation, reprinting, reuse of illustrations, recitation, broadcasting, reproduction on microfilms or in any other physical way, and transmission or information storage and retrieval, electronic adaptation, computer software, or by similar or dissimilar methodology now known or hereafter developed.
The use of general descriptive names, registered names, trademarks, service marks, etc. in this publication does not imply, even in the absence of a specific statement, that such names are exempt from the relevant protective laws and regulations and therefore free for general use.
The publisher, the authors and the editors are safe to assume that the advice and information in this book are believed to be true and accurate at the date of publication. Neither the publisher nor the authors or the editors give a warranty, expressed or implied, with respect to the material contained herein or for any errors or omissions that may have been made. The publisher remains neutral with regard to jurisdictional claims in published maps and institutional affiliations.

This Springer imprint is published by the registered company Springer Nature Switzerland AG
The registered company address is: Gewerbestrasse 11, 6330 Cham, Switzerland

If disposing of this product, please recycle the paper.

Foreword

This is the final book of six I am editing in the *Machine Translation: Technologies and Applications* series (https://www.springer.com/series/15798) which I agreed to co-edit back in 2016, following an approach by Jolanda Voogd, my Springer contact at the *Machine Translation* journal.

When I first proposed this to Springer, the series was to "tackle prominent issues in MT today at a depth which will allow these books to reflect the current state-of-the-art, while simultaneously standing the test of time". Fortunately, Springer agreed to commission the series, and to date, the first five books in the series are as follows:

- *Translation Quality Assessment* (2018): Joss Moorkens, Sheila Castilho, Federico Gaspari, & Stephen Doherty (eds.)
- *A Neurolinguistic Solution to Ambiguity and Complexity in Machine Translation* (2018): Bernard Scott
- *Explorations in Empirical Translation Process Research* (2021): Michael Carl (ed.)
- *Towards Responsible Machine Translation: Ethical and Legal Considerations in Machine Translation* (2023): Helena Moniz & Carla Parra Escartín (eds.)
- *Sign Language Machine Translation* (2024): Andy Way, Lorraine Leeson and Dimitar Shterionov (eds.) (forthcoming)

The first four of these have come out, and at the time of writing, have already been accessed over 85,000 times. By far the most popular to date is the book by my colleagues Joss Moorkens et al., as whatever paradigm is in vogue—as well as whether MT is being used at all—the ability to tell whether translations are any good is crucial. Importantly, being able to inspect translations and report precisely on why they are insufficient is of great importance. Bud Scott's book came out around the same time, and as it inspects the neurolinguistic assumptions underpinning the Logos MT system, it remains relevant for today's almost entirely neural models. Michael Carl's volume inspects MT as part of the overall translation process. Given the ubiquity of MT nowadays, and knowing first-hand from my days in industry what a disruptive technology MT is, the papers in this book will remain relevant for years. The topics tackled in Helena and Carla's book are hugely important, increasingly so, as we see a greater focus on how AI applications can be built and deployed 'for good'. Since the book was published, we have seen the publication of the AI Act in Europe, with more vigilance and no doubt legislation to come globally, so the writing of this book was hugely prescient, as we all seek to see AI being used responsibly.

The 5th book on sign language MT is about to come out. It is all too easy these days for people to talk about AI applications being 'solved problems', but today's technology is so dependent on data being available that it is only when one works on a topic where data is extremely scarce that one can see immediately how foolish such claims are. As well as proponents of sign language MT, this volume is likely to appeal equally to anyone working in low-resource applications, especially in language processing.

It is easy to see a common thread between the previous five books, but at first glance, the reader may struggle to see the significance of this book in the series, so let us try to explain. Firstly, all three editors have a profound interest in MT. Secondly, many of the breakthroughs in AI have come initially from MT; a brief literature review will reveal that many groundbreaking papers, such as the well-known Seq2Seq paper by Sutskever et al., the Encoder–Decoder paper from Cho et al., the famous Attention paper by Bahdanau et al., and the hugely influential Transformer paper by Vaswani et al. were either proposed directly for MT or extensively tested on it. This highlights the critical role that the MT community plays in shaping next-generation solutions in the field. The complexity of MT, which involves handling (at least) two different languages with independent data distributions and bridging the space between them, is so high that investigating these issues can provide valuable inspiration for addressing other NLP challenges. Moreover, while large datasets with billions of tokens are now common, historically, MT was one of the few ML tasks with millions of training samples (the English–French corpus of WMT in 2014, a decade ago, includes over 40 million parallel sentences according to https://aclanthology.org/W14-3302.pdf), while back then other NLP datasets only provided a few hundred or thousands of samples. Improvements in quality have led to an increased focus on efficiency and effectiveness within the MT community. For example, *n*-gram language models, which were large even in earlier times (with billions of parameters according to https://aclanthology.org/J92-4003.pdf), required efficient implementations, motivating the community to prioritize performance and efficiency early on.

Accordingly, this volume aims to move beyond translational detail and explore language understanding and network training much more from an engineering standpoint. In this volume, we dissect the underlying architecture of AI models, viewing them through an engineering lens. Availing of the range of techniques expounded in this book will benefit any practitioners using LLMs, including multilingual versions for translation.

At the time the series was commissioned, I boldly informed Springer that "there are, in fact, very few books published on Machine Translation in particular, or Translation more generally. Accordingly, there is no competition for the intended series in this space." MT is not a solved problem, so it is terrific to see that there have been many books published on the topic recently, including:

- *Translation Project Management* (2022): Callum Walker
- *The Human Translator in the 2020s* (2023): Gary Massey, Elsa Huertas-Barros, David Katan (eds.)
- *Translation Tools and Technologies* (2023): Andrew Rothwell, Joss Moorkens, Maria Fernández Parra, Joanna Drugan, Frank Austermuehl
- *Using Technologies for Creative-Text Translation* (2023): James Luke Hadley, Kristiina Taivalkoski-Shilov, Carlos S. C. Teixeira, Antonio Toral (eds.)
- *Translation Ethics* (2023): Joseph Lambert
- *Computer-Assisted Literary Translation* (2023): Andrew Rothwell, Andy Way, Roy Youdale (eds.)
- *Automating Translation* (2024): Joss Moorkens, Andy Way, Séamus Lankford (forthcoming)
- *Localization in Translation* (2024): Miguel A. Jiménez-Crespo
- *Working as a Professional Translator* (2024): JC Penet

Hopefully, then, more and more people will be inspired by what they have read and will want to work in this most interesting of all applications ... and this book will help them do so more efficiently!

Finally, I'd particularly like to thank Springer for trusting me in a number of roles for the past 20 years: as Book Review Editor and subsequently Editor of the *Machine Translation* journal, which ran until 2021; and thereafter with the stewardship of this series. Hopefully the journal and these books have demonstrated some value to the MT community. I'd also like to thank my *DCU* colleagues Joss, Sheila and Federico, ex- *DCU* colleagues Carla, Stephen and Dimi, my host in the beautiful city of Lisbon and current *EAMT* president Helena, my *ADAPT* colleague Lorraine, and my old EBMT colleague Michael, as well as Bud, for all their hard work in putting these volumes together. As they will be able to testify (albeit with gritted teeth!), I have read every word, corrected a few and made changes to others, but they were very much responsible for getting together the content that you have had the privilege of reading. Around ten years ago, when I returned to *DCU* to set up *ADAPT*, I was very fortunate that my colleague at the time Qun Liu had assembled a fantastic array of Ph.D. talent, which he very generously offered to me to co-supervise. One of those students was Peyman, who a few years after graduating I convinced to edit a volume on AI approaches to MT and NLP. No differently from how we was as a PhD student, together with Mehdi, he did this with boundless enthusiasm and extreme efficiency; I was again fortunate to be invited by them to help co-edit this volume, so thanks to both Peyman and Mehdi for the opportunity to work on this, my last book.

But my time is now up, and it's over to the younger generation, including Peyman and Mehdi, to pick up the gauntlet and take the subject forward, as I quietly slip away into retirement. With people like them and Helena at the helm, the field is in very safe hands indeed. Springer wish for this series to continue, so we all look forward to future volumes building on what has been produced to date.

Dublin, Ireland
July 2024

Andy Way

Preface

This book provides an in-depth investigation of the dynamic world of large language models (LLMs), which are revolutionizing various facets of research and application within the fields of deep learning, natural language understanding, medical image analysis, predicting the properties of proteins, understanding language acquisition, and many others. We have observed that LLMs not only represent a hot topic but signify a potentially fundamental paradigm shift, one which may lead to the dawn of a new era of research and so merits significant attention. These models pose unique challenges due to their scale, requiring unconventional training and data collection approaches that differ from traditional methods.

Beyond basic language processing, LLMs offer expanded capabilities that warrant further exploration. This broadens their utility and introduces additional challenges, particularly concerning their operation as intelligent agents. Questions of privacy, security, and ethical usage are paramount and highlight the distinct nature of these models compared to their predecessors. It is these differences and the need for a comprehensive understanding that inspired us to compile this book.

This year, following the third annual workshop on *Efficient Natural Language and Speech Processing*,[1] we recognized the opportunity to transform some of the workshop's content into a published format for the benefit of the broader community. We realized that researchers provide small but vital techniques to train, fine-tune, optimize, and accelerate LLMs. These incremental but crucial innovations often bypass the traditional publication process, where a theoretical method is enshrouded in a lengthy paper accompanied by an extensive literature review. Instead, today's researchers lean towards presenting engineering solutions and diving directly into the main techniques without much preliminary exposition. We felt it important that this evolving approach be documented in a book, and one which could appear as quickly as possible in this fast-moving area, so that other practitioners could benefit from the range of techniques contained herein. By doing so, we not only preserve the ingenuity of contemporary researchers but also provide a reliable resource for more traditional learners yet wish to access cutting-edge material. This ensures that

[1] https://neurips.cc/virtual/2023/workshop/66532

valuable practical insights are captured and made accessible to a wider audience. Additionally, by documenting these insights, we make sure that the engineering claims align with the scientific standards and expectations of the research community. This balance enhances the credibility of the techniques presented and fosters a more robust dialogue between developers of practical applications and researchers interested in adding to our theoretical understanding of how these models work, and how they can be made to work more effectively.

The main theme of this book is *efficiency* and the pivotal topic is “scale”. More specifically, in this volume, we aim to examine the reasons behind the substantial size of LLMs, investigate the intricacies of their design and the consequent implications. We will discuss the formidable challenges they pose, as well as the unprecedented opportunities they offer. The discussion extends to various technical considerations such as model training, selection of data sets, and the architecture of LLMs. In the first introductory chapter, we lay out a roadmap for the journey ahead, detailing what readers can expect from each subsequent section and chapter of the book. Additionally, we provide the basic fundamentals necessary to understand LLMs, ensuring that regardless of their prior knowledge, readers have a solid foundation from which to explore more advanced concepts throughout the book. The following chapters will not shy away from (sometimes quite deep) detail, as dissecting the intricacies of LLMs is critical to aid understanding of this new paradigm.

This book adopts a direct and focused approach, intentionally putting slightly less weight on traditional literature review and background exposition to immediately engage with proposed techniques and relevant details. It is crafted for a diverse audience, including students, practitioners, and both junior and senior scientists and engineers, whether in academia or industry. Designed to cater to a broad spectrum of readers, it ranges from those seeking detailed technical insights into the workings of LLMs to those interested in understanding the broader implications of these models in practical and theoretical contexts. By maintaining a careful balance between in-depth technical explanations and overarching discussions, this book aims to be both accessible and valuable, swiftly moving to enrich the reader’s knowledge and appreciation of LLMs without presupposing extensive prior knowledge of the field.

Toronto, Canada — Peyman Passban
Toronto, Canada — Mehdi Rezagholizadeh
Dublin, Ireland — Andy Way
May 2024

Acknowledgements

Andy has thanked a number of people in the Preface, but in addition, we would like to thank all of the contributors to this volume for delivering content and reacting to our queries very speedily. In addition, we thank our anonymous reviewers for their comments which improved the book. Finally, we thank members of the scientific community for providing testimonials on the contents of this volume.

Toronto, Canada
Toronto, Canada
Dublin, Ireland
May 2024

Peyman Passban
Mehdi Rezagholizadeh
Andy Way

Contents

Acronyms

A-V	Audio-Visual
AAM	Additive Angular Margin
ACC	Unsupervised Clustering Accuracy
AHC	Agglomerative Hierarchical Clustering
AMI	Adjusted Mutual Information
ASP	Attention Statistics Pooling
BERT	Bidirectional Encoder Representations from Transformers
BLEU	BiLingual Evaluation Understudy
BLSTM	Bidirectional Long Short-Term Memory
BPC	Bit Per Character
CA	Cross-Attention
CE	Cross Entropy
CHS	Calinski-Harabasz Score
CIMC	Contrastive Information Maximization Clustering
CL	Classification
CMT	Cross-Modal Transformer
CNN	Convolutional Neural Network
CoLA	The Corpus of Linguistic Acceptability
DBS	Davies-Bouldin Score
DEC	Deep Embedded Clustering
ECAPA	Emphasized Channel Attention, Propagation and Aggregation
EER	Equal Error Rate
FFN	Feed-Forward Network
FFT	Fast Fourier Transform
FLOPS	Floating Point Operations Per Second
FMI	Fowlkes-Mallows Index
FP	Floating Point
GELU	Gaussian Error Linear Unit
GLUE	General Language Understanding Evaluation
GMM	Gaussian Mixture Model
GPT	Generative Pre-Trained Transformer

GPU	Graphics Processing Unit
GSM8k	Grade-School Math
IMSAT	Information Maximizing Self-Augmented Training
INT	Integer
JCA	Joint Cross-Attention
KD	Knowledge Distillation
KL	Kullback Leibler
KNN	N-Nearest Neighbors
KU	Kaiming Uniform
KV	Key/Value
KronA	Kronecker Adapter
LM	Language Model
LLMs	Large Language Models
LRU	Linear Recurrent Unit
LR	Learning Rate
LSSM	Linear State Space Model
LSTM	Long Short-Term Memory
LoDA	Low-Dimensional Adaptation
LoRA	Low-Rank Adaptation
MEGA	Moving Average Equipped Gated Attention
MFB	Mel FilterBank
MFCCs	Mel-Frequency Cepstral Coefficients
MI	Mutual Information
MLM	Masked Language Modelling
MLP	Multi Layer Perceptron
MNLI	Multi-Genre Natural Language Inference
MOPs	Memory Operations
MPT	Mosaic Pre-Trained Transformer
MRPC	Microsoft Research Paraphrase Corpus
MSE	Mean Squared Error
NLP	Natural Language Processing
NMI	Normalized Mutual Information
PA	Parallel Adapter
PEFT	Parameter-Efficient Fine-Tuning
PKD	Patient Knowledge Distillation
PLM	Pre-trained Language Model
QNLI	Question-answering Natural Language Inference
QQP	Quora Question Pairs
R-LoDA	Low-Dimensional Adaptation based on low-Rank approximated weight
RIR	Room Impulse Response
RJCA	Recursive Joint Cross-Attention
RMS	Root Mean Square
RTE	Recognizing Textual Entailment
ReLU	Rectified Linear Unit
S-LoDA	Low-Dimensional Adaptation based on Sparsified weight

SAT	Self-Augmented Training
SA	Self-Attention
SFT	Supervised Fine-Tuning
SOM	Self-Organizing Maps
SST-2	The Stanford Sentiment Treebank
STS-B	Semantic Textual Similarity Benchmark
SV	Speaker Verification
SiLU	Sigmoid Linear Unit
SupCon	Supervised Contrastive Loss
T5	Text-to-Text Transfer Transformer
TCNCA	Temporal Convolutional Network with Chunked Attention
TCN	Temporal Convolutional Network
TDNN	Time Delay Neural Network
TPU	Tensor Processing Unit
VAE	Variational Autoencoder
VAT	Virtual Adversarial Training
VMT	Virtual Mixup Training
WER	Word Error Rate

Part I
Fundamentals

Chapter 1
Introduction and Fundamentals

Peyman Passban, Mehdi Rezagholizadeh, and Andy Way

Abstract In this chapter, we explain the intricacies of language modelling, focusing on the evolution from statistical models to the sophisticated large language models (LLMs) that dominate the field today. We explore the transition from n-gram models to neural network-based approaches, highlighting key advancements such as *Word2Vec*, *ELMo*, *BERT*, and the *Transformer* architectures. The chapter emphasizes the significance of scale in LLMs, discussing how increased model size enhances their capabilities, including context understanding and emergent behaviour. We also address the challenges associated with pre-training and fine-tuning these models, providing insights into data requirements, structural adaptations, and the implications of scaling laws. Finally, we examine the impact of model scale on predictive mechanisms and the tendency for hallucinations, proposing potential solutions to mitigate these issues.

1.1 Language Modelling

Language modelling is simply a task of computationally modelling a language, which is achieved through predicting the next word or sequence of words in a piece of text. This foundational technique underpins a wide range of applications, including text generation and speech recognition. Though the concept of language modelling is not new, it has undergone significant evolution from its origins. Initially, the field relied on statistical models such as *n*-gram models (Pauls and Klein 2011), which used fixed-size sequences of words to predict the likelihood of subsequent text continuations. These early models laid the groundwork for the more complex approaches. However, they were limited in their ability to capture contextual nuances. Thus, the evolution continued with the introduction of neural network-based approaches, notably through

P. Passban (✉) · A. Way
ADAPT Centre, School of Computing, Dublin City University, Dublin, Ireland
e-mail: peyman.passban@adaptcentre.ie

M. Rezagholizadeh
Noah's Ark Lab, Huawei, Markham, ON, Canada

© The Author(s), under exclusive license to Springer Nature Switzerland AG 2025
P. Passban et al. (eds.), *Enhancing LLM Performance*, Machine Translation: Technologies and Applications 7, https://doi.org/10.1007/978-3-031-85747-8_1

the pioneering work of Bengio et al. (2003), Collobert and Weston (2008), and technologies like Word2Vec (Mikolov et al. 2013).

More neural models have been proposed with different flavours and functionalities, such as recurrent models (Mikolov et al. 2010) and memory-based architectures (Sundermeyer et al. 2012) to improve context retention. Note too that by their very nature, these models dealt with just a single language, and so were restricted to monolingual applications. However, simply adding words and phrases from different languages allowed language models (LMs) and today's large language models (LLMs) to become multilingual (MLLMs), so that the same basic approach can be used for multilingual applications like machine translation (MT).

Building upon this progress, Bahdanau et al. (2015) had introduced the concept of "*attention*" to the recurrent architectures, where "a natural line of investigation was to see whether attention could do most of the heavy lifting" of context learning by itself (Moorkens et al. 2024, Chap. 4). That also enabled innovations such as bidirectional models (Rahman et al. 2021). The introduction of more advanced context-based models like ELMo (Peters et al. 2018) and BERT (Devlin et al. 2019) further revolutionized the field by enabling even deeper understanding and processing of linguistic context. The most significant breakthrough came with the development of the Transformer architecture (Vaswani et al. 2017) (see Chap. 2, Sect. 2.2.1 for a technical description). Its natural capability to parallelize computations, the self-attention mechanism, and other unique features such as tensor normalization modules allowed models to weight the importance of different words within a sentence and learn more about the context, all while allowing the use of much simpler feed-forward neural networks.

LLMs are firmly rooted (Brown et al. 2020a) in the Transformer paradigm, basically being deep, large Transformers. However, recent discussions surrounding LLMs signal a transformative leap rather than a simple extension of these models, which is the main theme of this chapter. From a different point of view, a more balanced perspective is also conceivable, as discussed by Way (2024) and further explored in Chap. 11 of this volume.

LLMs differ fundamentally not merely due to their size but because of the emergence of distinct features that become prominent at such scales. Their vast capacity allows for an extensive encapsulation of human knowledge and context, markedly setting them apart from their predecessors. This capacity enables extraordinary capabilities that were previously unattainable, prompting researchers to consider not just the models themselves but also the ecosystems in which they operate. This holistic view highlights the unique nature of LLMs, as they interact with and adapt to their computational and data environments in ways that smaller models cannot.

One development that emerged with the increasing scale and size of models was the concept of pre-training on vast datasets (Devlin et al. 2019). This pre-training allows the models to function as general-purpose tools, capable of adapting to a wide range of possible future applications. When specific needs arise, these models can be fine-tuned with relatively little effort to become specialized tools tailored to precise tasks. This process enhances their versatility and efficacy in various contexts.

To understand the training and fine-tuning of these models, it is essential to recognize that the mathematical principles underlying the training of LLMs are akin to those of earlier models, though applied at a much larger scale. Typically, an LLM operates by predicting the next word in a sequence based on the history of previous tokens. The mathematical formulation for next-word prediction can be simply expressed as in Eq. (1.1):

$$P(w_{n+1}|w_1, w_2, ..., w_n) = Softmax(f(w_1, w_2, ..., w_n)) \tag{1.1}$$

where w_{n+1} is the next word to be predicted and w_1, ..., w_n denote the previous words in the sequence. The function f embodies the transformations computed by the model's architecture to predict the probability of w_{n+1} facilitated by a *Softmax* layer that normalizes the outputs to ensure they represent valid probabilities.

The concept may seem simple–it is just next-word prediction–yet there are many debates about whether this is sufficient for sophisticated language understanding. However, experimental results demonstrate that this straightforward approach works quite effectively. Again, the secret lies in the scale. Through this technique, the model has a chance to predict the probability of billions or even trillions of words, which helps it form an internal representation of its world and capture the nuances of language with remarkable accuracy on the whole, given the simplicity of the approach.

It should be noted that it is not entirely accurate to claim that next-word prediction is the only method for training language models. Researchers have proposed various alternatives, such as masked language modelling, next sentence prediction, and permutation language modelling (Yang et al. 2019). These different approaches have significantly contributed to the community's understanding and advancement of language modelling techniques. Nevertheless, it must be acknowledged that, at present, next-word prediction has been demonstrated to be the most effective method among these options for achieving sophisticated language understanding capabilities.

Pre-training neural models at scale is extremely challenging and demands specialized strategies. These challenges and potential remedies will be discussed in subsequent chapters. The pre-training phase must ensure that models are exposed to a vast amount of data, learn from this data effectively, avoid catastrophic forgetting, and utilize resources efficiently. Achieving these objectives is crucial for developing high-quality models capable of supporting the backbone of language understanding (and other) applications.

However, these generic models might not be sufficiently accurate or domain-specific for certain tasks, necessitating customized versions. This is where fine-tuning comes into play. Fine-tuning leverages the extensive language understanding of pre-trained LLMs and adapts them for specialized applications using relatively small labeled datasets. This process enhances the model's performance on specific tasks while retaining the general knowledge acquired during pre-training.

More formally, fine-tuning adapts the LLM to perform specific tasks using labeled datasets. This involves supervised learning where the model parameters are adjusted to minimize the loss for a particular task, such as sentiment analysis, MT, or question answering.

For auto-regressive (AR) language modelling, the objective is to maximize the likelihood of the next token given the previous tokens (very similar to the pre-training phase), as formulated in Eq. (1.2):

$$\mathcal{L}_{AR} = -\sum_{n=1}^{N} \log P(w_n | w_{<n}; \theta) \tag{1.2}$$

where θ is the set of model parameters and the prediction of w_n relies on its preceding words. For masked language modelling (MLM) the criterion is slightly different that a masked token is predicted based on the context information gathered from unmasked, context words ($\hat{m}$), as shown in (1.3):

$$\mathcal{L}_{MLM} = -\sum_{m} \log P(w_m | w_{\hat{m}}; \theta) \tag{1.3}$$

A model can also be optimized on a task-specific dataset during fine-tuning, as shown in (1.4):

$$\mathcal{L}_{CL} = -\sum_{(x,y)} \log P(y|x; \theta) \tag{1.4}$$

where x is the input (a sequence of representations of words) and y is an associated label defined by that specific classification task. Model fine-tuning is not limited to these methods, and any strategy that can inject in-domain knowledge to LLMs could be considered as a fine-tuning method. Radford et al. (2018) illustrated this concept very well from the perspective of a range of different tasks, so is worth consulting for newcomers to the field.

1.2 The Scale Factor in LLMs

When discussing the scale factor in LLMs, it is essential to understand how it fundamentally influences their capabilities and behaviours. The term "large" does not only refer to the scale but consists of a range of aspects that collectively enhance the model's performance and applicability. To illustrate, GPT-3 has 175 billion parameters (Brown et al. 2020a), or PaLM is even a larger model with 540 billion parameters, 3x the size of GPT-3 (Chowdhery et al. 2023a), and these are not the largest models on the market; Google have already released a solution with a capacity to train the first *trillion*-parameter model (Fedus et al. 2021).

To put these advancements into perspective, it is useful to compare them with earlier models. The BERT model, that was not so long ago referred to as the state-of-the-art model for many natural language processing (NLP) tasks, contains 330 million parameters (Devlin et al. 2019), while GPT-2, another significant model, has 1.5 billion parameters (Radford et al. 2019). Although these models were revolutionary at their time, their parameter count pales in comparison to those of recent models. These figures are not just substantial statistics; they represent the extensive capacity of LLMs to learn and generate information. The enormous number of parameters enables LLMs to absorb so much information such that leads to emergent behaviours that smaller models cannot exhibit. Nevertheless, not all behaviour is entirely positive. In Chap. 11, we discuss related issues of the negative impact of huge electricity consumption and CO_2 emissions, as well as approaches like knowledge distillation used in Chap. 8 to reduce these effects without significantly adverse effects on performance.

The scale of LLMs extends beyond the number of parameters. For instance, PaLM is pre-trained on a staggering number of 6144 TPU v4 chips (Chowdhery et al. 2023a), highlighting the computational resources required for such models, or fine-tuning of models like Flan-PaLM involved 473 diverse datasets (Chung et al. 2024), underscoring the breadth of data used for these models. Even we see the footprint of larger scales in the inputs to these models. OpenAI released GPT-4 Turbo with a 128K context window and Anthropic's Claude 2.1 has a 200K context window.[1] The exponential increase in the size and complexity of LLMs reflects a broader trend where models demonstrate superior performance more and more across a wide range of tasks.

At this point, it is evident that the first "L" in LLMs plays a key role but understanding why is crucial. The primary reason is the relationship between scale and performance. As discussed, when the scale of these models increases, researchers observe emergent behaviours that significantly enhance their capabilities. One of the most important findings is that LLMs exhibit improved context understanding as they grow larger. Brown et al. (2020b) visualized this concept very well, showing that the larger the GPT model, the higher the accuracy. Moreover, they demonstrated that larger models perform considerably better on zero-, one-, and few-shot learning tasks, meaning that the models find context to be beneficial.

Recent research has shown that smaller models often suffer from saturation, where their performance drops and plateaus after a certain point in training. To overcome this, models must exceed a certain threshold in size to exhibit certain behaviours (Godey et al. 2024). A tangible example of this phenomenon can be observed when LLMs complete various types of text sentences. Not only do these sentences appear fluent and grammatically correct–an achievement in itself compared to conventional models–but they also provide semantically and factually accurate answers to complex questions. This dual capability (fluency and correctness) is a clear example of emergent behaviour. These capabilities are particularly evident in complex applications

[1] https://www.anthropic.com/news/claude-2-1.

such as coding,[2] where the model's output is both syntactically correct and contextually appropriate (note that this behaviour clearly goes beyond next word prediction or sentence completion).

Another notable emergent behaviour in LLMs is step-by-step reasoning or the "chain-of-thought" (Wei et al. 2022) process. This concept refers to the model's ability to break down complex problems into smaller, manageable steps and provide a detailed explanation of the reasoning process involved. As these models store vast amounts of information within their internal structures, to some extent they are capable of performing limited but significant forms of reasoning. This chain-of-thought reasoning is remarkable because it enables the model to handle tasks that require logical progression and problem-solving skills. Furthermore, it allows LLMs to provide explanations for their answers, making them more transparent and interpretable. Users can follow the model's 'thought process', understand the 'rationale' behind its conclusions, and assess the validity of its 'reasoning'.

1.3 Implications of the Scale Factor in LLMs

In the previous section, we showed that the scale of LLMs is not just a matter of having more parameters; it is a critical factor that enables them to perform tasks with unprecedented accuracy and sophistication. The emergence of advanced behaviours such as context understanding and chain-of-thought also demonstrate the importance of scale. However, this scale factor influences almost all aspects of LLMs, from model training, data ingestion, to many others.

1.3.1 Data Complications

The first significant impact of the scale is on data requirements. These models are exceptionally data-hungry, necessitating not only vast quantities of data but also diverse sets of data to ensure comprehensive training. To provide some context regarding the scale and variety of these datasets, consider the CommonCrawl dataset, C4, which has a size of 800GB (Raffel et al. 2020). Another CommonCrawl dataset, REALNEWs, is 120GB (Zellers et al. 2019). Wikipedia's dataset comprises 21GB of content, while OpenWebText includes 38GB of Reddit data.[3] These examples illustrate the complexity and magnitude of the data involved.

A few years ago, even transferring such massive datasets to a cloud environment was a challenging task, let alone pre-processing, tokenizing, and training models with them. Today, these tasks have become more feasible, yet they remain non-trivial. For instance, fine-tuning Flan-PaLM involved using 473 datasets, covering 146 unique

[2] https://github.com/OpenDevin/OpenDevin.

[3] https://skylion007.github.io/OpenWebTextCorpus/.

task categories and totaling 1836 tasks (Chowdhery et al. 2023). This process required extensive data cleaning, initiating numerous training cycles, collecting results, and continuously optimizing configurations.

Access to such large-scale data is rarely feasible across all domains (see Way (2024) and Chap. 11 for more on this). Even when large datasets are available, adjustments are often necessary to achieve ideal model performance. Research indicates that the ratio of different data sources directly influences model quality. Accordingly, getting this balance right is a complex engineering challenge rather than a straightforward data aggregation task. For example, "the influence of the source language in MT has long been known, but, a different kind of influence, namely that of the predominant language in the training set" (Way 2024), is starting to be clearly seen in the development of (M)LLMs (Moorkens et al. 2024; Naous et al. 2023). Thus, careful down- or up-sampling is often required to balance the data sources effectively (Clark et al. 2020). Studies have shown that removing data sources with high heterogeneity, such as web pages, has a more detrimental impact on LLM abilities than removing more homogeneous sources, such as academic corpora. Additionally, ratio might not be the only element to consider. The order of different data instances can also significantly affect learning outcomes. This concept, known as "curriculum learning", involves the adjustment of data proportions from different sources. The order in which data is presented plays a crucial role in in-context learning and can markedly influence the performance of LLMs (Liu et al. 2024).

1.3.1.1 Data Quality and Availability

Data complications are not only limited to technical aspects. Non-technical issues are also important, so we decided to create a dedicated section for them, rather than mixing them with technical matters or scattering them throughout the text. The previous discussion highlighted the significance of data applications and how their size and quality necessitate new approaches. However, accessing such large datasets is not always feasible. According to research[4] from the European Language Equality project (Rehm and Way 2023), most European languages are in danger of digital extinction, with some even facing actual extinction. In this scenario, LLMs might in fact prove effective, as their zero- and few-shot learning capabilities allows the use of less data to achieve more. They can even help generate synthetic data for these languages, but *is this always a viable solution*?

Thompson et al. (2024) recently demonstrated that a substantial amount of bilingual data mined for training has been machine-translated, which "raises serious concerns about training models such as multilingual large language models on both monolingual and bilingual data scraped from the web". Consequently, proper labeling and careful creation of these datasets are imperative. Another example noted by Way (2024) is that "Siddhant et al. (2022) cover 'the Next 1000 Languages in Multilingual Machine Translation', but despite this promising title, they demonstrate that

[4] https://european-language-equality.eu/.

much data on the web attributed to certain languages is wrongly ascribed". These examples reinforce the importance of the availability and quality of data, regardless of the technical complications.

The process of dataset curation and collection is also profoundly affected. To gather large datasets, a significant portion (if not all) of the textual data available on the web must be harvested. This approach can potentially raise serious issues concerning intellectual property infringements, unauthorized access to data, and data leaks, among other concerns. While huge datasets are crucial for training (M)LLMs, this aspect also demands careful consideration and attention from researchers. The implications of data usage must be managed with strict protocols to prevent breaches of privacy and ensure ethical compliance. We recommend the work of Strubell et al. (2019), Leins et al. (2020), Bender et al. (2021), Birhane et al. (2021), and Carlini et al. (2021), as well as the contents of the edited volume by Moniz H, Escartin (2023), all of which extensively investigate such issues, the latter from an MT perspective.

1.3.2 Structural and Architectural Adaptations

Data is not the only factor impacted by the large size of these language models; the internal components of these models must also be adapted to handle the increased complexity and scale effectively. For instance, layer normalization was introduced in the seminal Transformer paper to cope with variance shift across layers. This technique helps stabilize the training process by normalizing the inputs to each layer, thereby maintaining consistency in the training dynamics. Following the introduction of layer normalization, several other techniques have been proposed to extend and improve upon this concept. For example, batch normalization is widely used in various neural network architectures. However, it struggles with variable length data, making it less suitable for NLP tasks. Consequently, layer normalization is often preferred for its ability to handle variable input lengths more effectively (Ba et al. 2016; Xiong et al. 2020).

To address some of the limitations of layer normalization, RMSNorm was introduced as a faster alternative. RMSNorm simplifies the normalization process by using the root mean square (RMS) of the inputs instead of their mean and variance, leading to reduced computational overhead (Zhang and Sennrich 2019). Another significant advancement is DeepNorm, which aims to stabilize the training of deep Transformers. DeepNorm modifies the residual connections in the network, enabling Transformers to scale up to 1,000 layers without encountering the instabilities typically associated with deep networks (Wang et al. 2024). This innovation is crucial for training very large language models, which require deep architectures to capture complex language patterns.

The proposed post-layer normalization (post-LN) in the vanilla Transformer, is not optimal for very large models. Thus, Transformers with pre-layer normalization (pre-LN) are more stable during training (Zhao et al. 2023; Minaee et al. 2024). Despite pre-LN performing worse in some scenarios compared to post-LN, it is often adopted in LLMs due to its training stability. These various LN techniques and adaptations highlight the evolving strategies used to manage the complexities introduced by scaling up language models.

LN and similar aspects operate at the single-component level, but the interaction and integration of these components within a model are also significantly impacted by scale. A model with billions of parameters cannot fit into the memory of a single processor, necessitating specialized treatments. Handling such large volumes of data is a challenge that requires robust orchestration mechanisms. Motivated by these requirements, new techniques and solutions have been developed. Researchers have proposed data parallelism, which involves splitting the data across multiple processors while each processor runs the same model architecture. This approach allows the training of large models by distributing the workload of processing the data. Pipeline parallelism was introduced to distribute different layers of LLMs across multiple GPUs, enabling efficient utilization of hardware resources and reducing memory bottlenecks (Narayanan et al. 2019).

Tensor parallelism is another significant advancement. This technique involves splitting the model's parameters (tensors) across multiple devices, allowing each device to handle a portion of the model's operations. Under these parallelism strategies, methods such as ZeRO (Rajbhandari et al. 2020) were developed to optimize memory usage by partitioning the model states across data-parallel processes, reducing memory redundancy. Several frameworks and tools have also been built to support these parallelism techniques. DeepSpeed (Rasley et al. 2020), an optimization library, is one of them that provides efficient training of large models by incorporating ZeRO and other optimizations. JAX[5] is a library for high-performance numerical computing that offers automatic differentiation and accelerates research by enabling hardware-accelerated machine learning. BMTrain[6] is another training framework designed to handle large-scale models with efficient memory management. FastMoE (He et al. 2022), a framework for training models with mixtures of experts, allows efficient scaling and utilization of multiple GPUs. vLLM (Kwon et al. 2023), an inference-time optimization library was designed to reduce latency and improve throughput. Internal components have also been fine-tuned to work better with these strategies. Flash Attention, for example, is an efficient attention mechanism that reduces memory usage and computation time, making it suitable for large models (Dao et al. 2022). LoRA was proposed to reduce the number of parameters needed for fine-tuning, thereby making the training of large models more feasible and efficient (Hu et al. 2022). Beyond these techniques, researchers have revisited common training techniques such as knowledge distillation (Xu et al. 2024) to train better LLMs and also transfer the knowledge from them to smaller, more deployable versions, to benefit from them in real-world applications, as well as being kinder to the planet (see Chap. 11 for more on this).

The list of innovations in this field is extensive and continually growing, making it challenging to document them all comprehensively. However, the continuous

---

[5] https://github.com/google/maxtext.

[6] https://github.com/OpenBMB/BMTrain.

development of new techniques and frameworks demonstrates the rapid pace of advancements in the training and optimization of LLMs.

1.3.3 Predictive Mechanisms and Output Challenges

As previously discussed, data aspects, training methods, and internal mechanics of these models were significantly impacted by their scale. However, the implications of scaling extend beyond technical adjustments. The way researchers and practitioners think about these models has also evolved. One pivotal concept that emerged is the conversations around scaling laws. Given the immense resources and time required to train a model over several months, it is imperative to have some assurance of its eventual success. Scaling laws provide a mechanism to predict the performance of larger models, helping researchers select more scalable architectures and plan resource allocation (Isik et al. 2024). This predictive capability mitigates the risk of investing time and resources into training models with uncertain outcomes.

The large size of these models has also changed our expectations regarding their outputs. While we have extensively discussed inputs (data considerations) and internal components, we should also briefly mention the outputs of these models. LLMs possess extensive knowledge across a wide array of topics; however, the depth of this knowledge can vary, leading to the generation of inaccurate or misleading statements in some cases. These inaccuracies are often referred to as "hallucination" (Huang et al. 2023) or "confabulations" (Smith et al. 2023); these occur when models generate information that is plausible yet incorrect or not grounded in reality.

The tendency of these models to hallucinate might stem from their fundamental design as next-word predictors. Although this mechanism enables them to learn a vast amount of information, there are no explicit processes to teach them reasoning skills, making hallucinations a likely consequence. Addressing this issue requires grounding model outputs in reality and developing methods to filter out inaccuracies. Techniques such as fine-tuning on domain-specific data (Jeong 2024), using external knowledge bases (Gao et al. 2024), and implementing post-processing filters (Clusmann et al. 2023) are being explored to mitigate hallucinations and improve the reliability of model outputs.

1.4 Conclusion and Next Chapters

In this chapter, we have tried to cover the fundamentals of LLMs, specifically addressing the significance of the scale factor and its impact on these models and their surrounding environment. Our writing approach is slightly different than what is offered in scientific books; rather than providing exhaustive details, we present key concepts to a degree that ensures comprehension before moving on, and we intentionally selected this style for our book. We recognize that the field is exceptionally vibrant, with a vast number of papers and readily accessible material available to readers. Therefore, we focus on the essentials, allowing readers to refer to other sources for more comprehensive information if they wish. For those interested in further study, we recommend two invaluable resources: the works of Gao et al. (2024) and Zhao et al. (2023). These comprehensive surveys on LLMs provide an excellent foundation for researchers in this field.

In subsequent sections, we will concentrate on efficiency techniques, discussing how to run these models faster and more cost-effectively during inference, improve them during fine-tuning, and explore potential enhancements to their training or architecture. The book concludes with a section on efficiency techniques for smaller models. It is important to remember that LLMs are not the solution to every problem. Smaller models remain highly useful across various fields, and the principles of efficiency apply to them as well. Therefore, it is crucial to give due attention to both large and small models. Finally of course, even if in certain circumstances an LLM-based approach gives ostensibly 'better' results, if its 'reasoning' cannot be made absolutely explicit, many companies (especially multinationals operating in traditionally conservative sectors) will lean toward a less-performant but more explainable solution.

References

Ba J, Kiros JR, Hinton GE (2016) Layer normalization. ArXiv https://api.semanticscholar.org/CorpusID:8236317

Bahdanau D, Cho K, Bengio Y (2015) Neural machine translation by jointly learning to align and translate. In: 3rd international conference on learning representations, (ICLR)

Bender EM, Gebru T, McMillan-Major A, Shmitchell S (2021) On the dangers of stochastic parrots: can language models be too big? In: Proceedings of the 2021 ACM conference on fairness, accountability, and transparency, association for computer machinery - ACM. https://dl.acm.org/doi/10.1145/3442188.3445922

Bengio Y, Ducharme R, Vincent P, Janvin C (2003) A neural probabilistic language model. J Mach Learn Res 3:1137–1155. https://api.semanticscholar.org/CorpusID:221275765

Birhane A, Prabhu VU, Kahembwe E (2021) Multimodal datasets: misogyny, pornography, and malignant stereotypes. ArXiv arXiv:abs/2110.01963. https://api.semanticscholar.org/CorpusID:238354158

Brown T, Mann B, Ryder N, Subbiah M, Kaplan JD, Dhariwal P, Neelakantan A, Shyam P, Sastry G, Askell A, Agarwal S, Herbert-Voss A, Krueger G, Henighan T, Child R, Ramesh A, Ziegler D, Wu J, Winter C, Hesse C, Chen M, Sigler E, Litwin M, Gray S, Chess B, Clark J, Berner C, McCandlish S, Radford A, Sutskever I, Amodei D (2020a) Language models are few-shot learners. In: Advances in neural information processing systems, vol 33, pp 1877–1901. https://proceedings.neurips.cc/paper_files/paper/2020/file/1457c0d6bfcb4967418bfb8ac142f64a-Paper.pdf

Brown T, Mann B, Ryder N, Subbiah M, Kaplan JD, Dhariwal P, Neelakantan A, Shyam P, Sastry G, Askell A, Agarwal S, Herbert-Voss A, Krueger G, Henighan T, Child R, Ramesh A, Ziegler D, Wu J, Winter C, Hesse C, Chen M, Sigler E, Litwin M, Gray S, Chess B, Clark J, Berner C, McCandlish S, Radford A, Sutskever I, Amodei D (2020b) Language models are few-shot learners. In: Advances in neural information processing systems, vol 33, pp 1877–1901. https://proceedings.neurips.cc/paper_files/paper/2020/file/1457c0d6bfcb4967418bfb8ac142f64a-Paper.pdf

Carlini N, Tramèr F, Wallace E, Jagielski M, Herbert-Voss A, Lee K, Roberts A, Brown T, Song D, Erlingsson Ú, Oprea A, Raffel C (2021) Extracting training data from large language models. In: 30th USENIX security symposium (USENIX Security 21), USENIX Association, pp 2633–2650. https://www.usenix.org/conference/usenixsecurity21/presentation/carlini-extracting

Chowdhery A, Narang S, Devlin J, Bosma M, Mishra G, Roberts A, Barham P, Chung HW, Sutton C, Gehrmann S et al (2023a) Palm: scaling language modeling with pathways. J Mach Learn Res 1–113

Chowdhery A, Narang S, Devlin J, Bosma M, Mishra G, Roberts A, Barham P, Chung HW, Sutton C, Gehrmann S et al (2023b) Palm: scaling language modeling with pathways. J Mach Learn Res 24:1–113

Chung HW, Hou L, Longpre S, Zoph B, Tay Y, Fedus W, Li Y, Wang X, Dehghani M, Brahma S, Webson A, Gu SS, Dai Z, Suzgun M, Chen X, Chowdhery A, Castro-Ros A, Pellat M, Robinson K, Valter D, Narang S, Mishra G, Yu A, Zhao V, Huang Y, Dai A, Yu H, Petrov S, Chi EH, Dean J, Devlin J, Roberts A, Zhou D, Le QV, Wei J (2024) Scaling instruction-finetuned language models. J Mach Learn Res 1–53. http://jmlr.org/papers/v25/23-0870.html

Clark K, Luong MT, Le QV, Manning CD (2020) ELECTRA: pre-training text encoders as discriminators rather than generators. In: International conference on learning representations (ICLR). https://openreview.net/pdf?id=r1xMH1BtvB

Clusmann J, Kolbinger FR, Muti HS, Carrero ZI, Eckardt JN, Laleh NG, Löffler CML, Schwarzkopf SC, Unger M, Veldhuizen GP et al (2023) The future landscape of large language models in medicine. Commun Med 3(1):141

Collobert R, Weston J (2008) A unified architecture for natural language processing: deep neural networks with multitask learning. In: International conference on machine learning. https://api.semanticscholar.org/CorpusID:2617020

Dao T, Fu D, Ermon S, Rudra A, Ré C (2022) FlashAttention: fast and memory-efficient exact attention with IO-awareness. Adv Neural Inf Process Syst 35:16344–16359

Devlin J, Chang MW, Lee K, Toutanova K (2019) BERT: Pre-training of deep bidirectional transformers for language understanding. In: Proceedings of the 2019 conference of the North American chapter of the association for computational linguistics (ACL): human language technologies, vol 1 (Long and short papers), pp 4171–4186

Fedus W, Zoph B, Shazeer N (2021) Switch transformers: scaling to trillion parameter models with simple and efficient sparsity. arXiv preprint arXiv:2101.03961

Gao Y, Xiong Y, Gao X, Jia K, Pan J, Bi Y, Dai Y, Sun J, Wang M, Wang H (2024) Retrieval-augmented generation for large language models: a survey

Godey N, de la Clergerie É, Sagot B (2024) Why do small language models underperform? studying language model saturation via the softmax bottleneck. ArXiv https://api.semanticscholar.org/CorpusID:269042847

He J, Zhai J, Antunes T, Wang H, Luo F, Shi S, Li Q (2022) Fastermoe: modeling and optimizing training of large-scale dynamic pre-trained models. In: Proceedings of the 27th ACM SIGPLAN symposium on principles and practice of parallel programming, Seoul, Republic of Korea, pp 120–134. https://doi.org/10.1145/3503221.3508418

Hu EJ, yelong shen, Wallis P, Allen-Zhu Z, Li Y, Wang S, Wang L, Chen W (2022) LoRA: low-rank adaptation of large language models. In: International conference on learning representations (ICLR). https://openreview.net/forum?id=nZeVKeeFYf9

Huang L, Yu W, Ma W, Zhong W, Feng Z, Wang H, Chen Q, Peng W, Feng X, Qin B, Liu T (2023) A survey on hallucination in large language models: principles, taxonomy, challenges, and open questions. ArXiv arXiv:abs/2311.05232. https://api.semanticscholar.org/CorpusID:265067168

Isik B, Ponomareva N, Hazimeh H, Paparas D, Vassilvitskii S, Koyejo S (2024) Scaling laws for downstream task performance of large language models. ArXiv arXiv:abs/2402.04177. https://api.semanticscholar.org/CorpusID:267499809

Jeong C (2024) Fine-tuning and utilization methods of domain-specific llms. ArXiv arXiv:abs/2401.02981. https://api.semanticscholar.org/CorpusID:266844751

Kwon W, Li Z, Zhuang S, Sheng Y, Zheng L, Yu CH, Gonzalez JE, Zhang H, Stoica I (2023) Efficient memory management for large language model serving with pagedattention. In: Proceedings of the ACM SIGOPS 29th symposium on operating systems principles, Koblenz, Germany
Leins K, Lau JH, Baldwin T (2020) Give me convenience and give her death: Who should decide what uses of NLP are appropriate, and on what basis? In: Proceedings of the 58th annual meeting of the association for computational linguistics. Association for Computational Linguistics, Online, pp 2908–2913. https://aclanthology.org/2020.acl-main.261
Liu Y, Liu J, Shi X, Cheng Q, Lu W (2024) Let's learn step by step: enhancing in-context learning ability with curriculum learning. arXiv preprint arXiv:2402.10738
Mikolov T, Karafiát M, Burget L, Cernocký J, Khudanpur S (2010) Recurrent neural network based language model. Int Speech Commun Assoc (INTERSPEECH) 2:1045–1048
Mikolov T, Chen K, Corrado GS, Dean J (2013) Efficient estimation of word representations in vector space. In: International conference on learning representations (ICLR). https://api.semanticscholar.org/CorpusID:5959482
Minaee S, Mikolov T, Nikzad N, Chenaghlu M, Socher R, Amatriain X, Gao J (2024) Large language models: a survey. arXiv preprint arXiv:2402.06196
Moniz H, Escartín CP (eds) (2023) Towards responsible machine translation: Ethical and legal considerations in machine translation. Springer, Cham, Switzerland
Moorkens J, Way A, Lankford S (2024) Translation and automation. Routledge, Abingdon, Oxon., UK, forthcoming
Naous T, Ryan MJ, Xu W (2023) Having beer after prayer? measuring cultural bias in large language models. ArXiv arXiv:abs/2305.14456. https://api.semanticscholar.org/CorpusID:258865272
Narayanan D, Harlap A, Phanishayee A, Seshadri V, Devanur NR, Ganger GR, Gibbons PB, Zaharia M (2019) Pipedream: generalized pipeline parallelism for dnn training. In: Proceedings of the 27th ACM symposium on operating systems principles, pp 1–15
Pauls A, Klein D (2011) Faster and smaller n-gram language models. In: Proceedings of the 49th annual meeting of the association for computational linguistics: human language technologies, pp 258–267
Peters ME, Neumann M, Iyyer M, Gardner M, Clark C, Lee K, Zettlemoyer L (2018) Deep contextualized word representations. In: Proceedings of the 2018 conference of the North American chapter of the association for computational linguistics: human language technologies, vol 1 (Long papers), New Orleans, Louisiana, pp 2227–2237. https://aclanthology.org/N18-1202
Radford A, Narasimhan K, Salimans T, Sutskever I (2018) Improving language understanding by generative pre-training. OpenAI Blog https://www.openai.com/blog/language-unsupervised/
Radford A, Wu J, Child R, Luan D, Amodei D, Sutskever I et al (2019) Language models are unsupervised multitask learners, p 9
Raffel C, Shazeer N, Roberts A, Lee K, Narang S, Matena M, Zhou Y, Li W, Liu PJ (2020) Exploring the limits of transfer learning with a unified text-to-text transformer. J Mach Learn Res 5485–5551
Rahman MM, Watanobe Y, Nakamura K (2021) A bidirectional LSTM language model for code evaluation and repair. Symmetry 13:247
Rajbhandari S, Rasley J, Ruwase O, He Y (2020) Zero: memory optimizations toward training trillion parameter models. In: SC20: international conference for high performance computing. Networking, storage and analysis, pp 1–16
Rasley J, Rajbhandari S, Ruwase O, He Y (2020) Deepspeed: System optimizations enable training deep learning models with over 100 billion parameters. In: Proceedings of the 26th ACM SIGKDD international conference on knowledge discovery & data mining, pp 3505–3506
Rehm G, Way A (eds) (2023) European language equality. Springer, Cham, Switzerland
Siddhant A, Bapna A, Firat O, Cao Y, Chen MX, Caswell I, Garcia X (2022) Towards the next 1000 languages in multilingual machine translation: exploring the synergy between supervised and self-supervised learning. arXiv preprint arXiv:2201.03110
Smith AL, Greaves F, Panch T (2023) Hallucination or confabulation? neuroanatomy as metaphor in large language models. PLOS Digit Health 2(11):e0000388

Strubell E, Ganesh A, McCallum A (2019) Energy and policy considerations for deep learning in NLP. In: Proceedings of the 57th annual meeting of the association for computational linguistics. Association for Computational Linguistics, Florence, Italy, pp 3645–3650. https://aclanthology.org/P19-1355

Sundermeyer M, Schlüter R, Ney H (2012) LSTM neural networks for language modeling. Int Speech Commun Assoc (INTERSPEECH) 2012:194–197

Thompson B, Dhaliwal MP, Frisch P, Domhan T, Federico M (2024) A shocking amount of the web is machine translated: insights from multi-way parallelism. ArXiv https://api.semanticscholar.org/CorpusID:266933239

Vaswani A, Shazeer NM, Parmar N, Uszkoreit J, Jones L, Gomez AN, Kaiser L, Polosukhin I (2017) Attention is all you need. In: Neural information processing systems. https://api.semanticscholar.org/CorpusID:13756489

Wang H, Ma S, Dong L, Huang S, Zhang D, Wei F (2024) Deepnet: scaling transformers to 1,000 layers. IEEE Trans Pattern Anal Mach Intell

Way A (2024) What does the future hold for translation technologies in society? In: Baumgarten S, Tieber M (eds) Routledge handbook of translation technology and society. Routledge, Abingdon. Oxon, UK

Wei J, Wang X, Schuurmans D, Bosma M, ichter b, Xia F, Chi E, Le QV, Zhou D (2022) Chain-of-thought prompting elicits reasoning in large language models. In: Advances in neural information processing systems, pp 24824–24837. https://proceedings.neurips.cc/paper_files/paper/2022/file/9d5609613524ecf4f15af0f7b31abca4-Paper-Conference.pdf

Xiong R, Yang Y, He D, Zheng K, Zheng S, Xing C, Zhang H, Lan Y, Wang L, Liu T (2020) On layer normalization in the transformer architecture. In: Proceedings of the 37th international conference on machine learning, pp 10524–10533. https://proceedings.mlr.press/v119/xiong20b.html

Xu X, Li M, Tao C, Shen T, Cheng R, Li J, Xu C, Tao D, Zhou T (2024) A survey on knowledge distillation of large language models. ArXiv arXiv:abs/2402.13116. https://api.semanticscholar.org/CorpusID:267760021

Yang Z, Dai Z, Yang Y, Carbonell J, Salakhutdinov RR, Le QV (2019) Xlnet: Generalized autoregressive pretraining for language understanding. In: Advances in neural information processing systems, vol 32

Zellers R, Holtzman A, Rashkin H, Bisk Y, Farhadi A, Roesner F, Choi Y (2019) Defending against neural fake news. In: Advances in neural information processing systems, vol 32

Zhang B, Sennrich R (2019) Root mean square layer normalization. In: Advances in neural information processing systems, vol 32, Vancouver, Canada. https://openreview.net/references/pdf?id=S1qBAf6rr

Zhao WX, Zhou K, Li J, Tang T, Wang X, Hou Y, Min Y, Zhang B, Zhang J, Dong Z et al (2023) A survey of large language models. arXiv preprint arXiv:2303.18223

Part II
Inference Time Efficiency

Chapter 2
SPEED: Speculative Pipelined Execution for Efficient Decoding

Coleman Hooper, Sehoon Kim, Hiva Mohammadzadeh, Hasan Genc, Kurt Keutzer, Amir Gholami, and Yakun Sophia Shao

Abstract Generative LLMs based on the Transformer architecture have recently emerged as a dominant foundation model for a wide range of Natural Language Processing tasks. Nevertheless, their application in real-time scenarios has been highly restricted because of the significant inference latency associated with these models. This is particularly pronounced due to the autoregressive nature of generative LLM inference, where tokens are generated sequentially since each token depends on all previous output tokens. It is therefore challenging to achieve any token-level parallelism, making inference extremely memory-bound. In this work, we propose SPEED, which improves inference efficiency by speculatively executing multiple future tokens in parallel with the current token using predicted values based on early-layer hidden states. For Transformer decoders which employ parameter sharing, the memory operations for the tokens executing in parallel can be amortized, which allows us to accelerate generative LLM inference. We demonstrate the efficiency of our method in terms of latency reduction relative to model accuracy and demonstrate how speculation allows for training deeper decoders with parameter sharing with minimal runtime overhead.

2.1 Introduction

The Transformer neural network architecture has recently revolutionized NLP, providing massive accuracy gains across a range of tasks (Devlin et al. 2019; Brown et al. 2020). In particular, there has been growing interest in applying Transformer decoders for generative tasks (Radford et al. 2019, Raffel et al. 2020). Unlike Transformer encoders which can process an entire input sequence in parallel, Transformer decoders must be applied autoregressively at inference time as each input token depends on the output classification for the previous token. This means that they exhibit low arithmetic intensity and are typically memory bandwidth-bound (Park

---

C. Hooper (✉) · S. Kim · H. Mohammadzadeh · H. Genc · K. Keutzer · A. Gholami ·
Y. Sophia Shao
University of California, Berkeley, USA
e-mail: chooper@berkeley.edu

© The Author(s), under exclusive license to Springer Nature Switzerland AG 2025
P. Passban et al. (eds.), *Enhancing LLM Performance*, Machine Translation: Technologies and Applications 7, https://doi.org/10.1007/978-3-031-85747-8_2

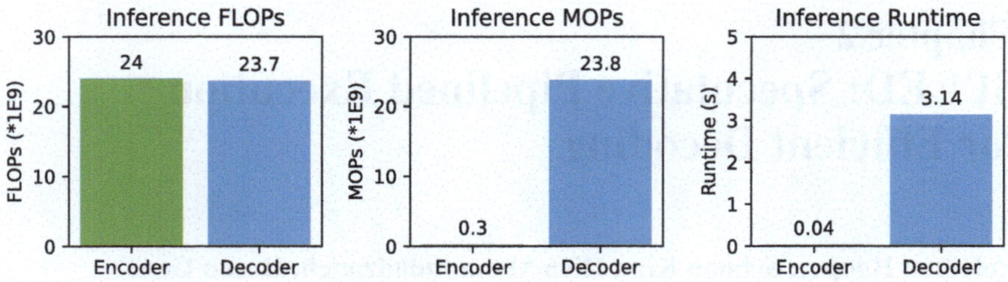

Fig. 2.1 From left to right: Floating-point operations (FLOPs), memory operations (MOPs), and end-to-end runtime on an A5000 GPU for an encoder-only model processing a sequence of 128 tokens in parallel, and for a decoder-only model generating 128 tokens autoregressively. The encoder and decoder model architecture parameters were set as the parameters from the T5-Base decoder-only model, which are provided for reference in Sect. 2.4.1. Although encoder and decoder inference have a similar number of FLOPs, the latency of autoregressive inference with a Transformer decoder is much greater due to the larger number of memory operations required. Profiling was performed using the Huggingface Transformers library (Wolf et al. 2020)

et al. 2020, Kim et al. 2023). For small batch sizes (as is typical for edge deployment scenarios (Schuster et al. 2022)), it is extremely difficult to achieve any parallelism. As shown in Fig. 2.1, the limited parallelism leads to increased runtime with Transformer decoders relative to encoders, even for the same sequence length (Park et al. 2020). In order to accelerate memory bandwidth-bound decoder inference, we must reduce the number of memory operations required.

In this chapter, we aim to reduce the latency of memory bandwidth-bound decoder inference by employing *speculative execution* in order to process tokens at different positions in the sequence *in parallel*. When employing speculative execution, the forward passes for future tokens are started using speculative output values from earlier tokens. By starting future tokens, we can process them *in parallel* with finishing the forward passes for earlier tokens. If a prediction is later found to be wrong, we must invalidate all future inferences that were started based on the speculative output value from the incorrect prediction. By still following all iterations through to completion, we can ensure that full model accuracy is maintained.

On its own, speculative execution would not lead to performance benefits within a single network. As shown in Fig. 2.2b, different tokens in the sequence would need to be processed by different layers in the network at the same time, meaning that the number of memory operations (MOPs) required for performing inference would not be reduced (even assuming perfect prediction). Additionally, to support inference on low-resource edge devices, it is crucial to reduce the model's memory footprint. *Parameter sharing* is a common method for model compression in Transformer networks (Lan et al. 2020, Reid et al. 2021). However, although it reduces the size of the network, parameter sharing does not typically provide significant speedup as the standard computation must still be performed for all layers in the network. Even if the inference is memory-bound, parameter sharing only reduces the number of MOPs required if the entire model fits in local cache memory, which is restrictive and hardware-dependent.

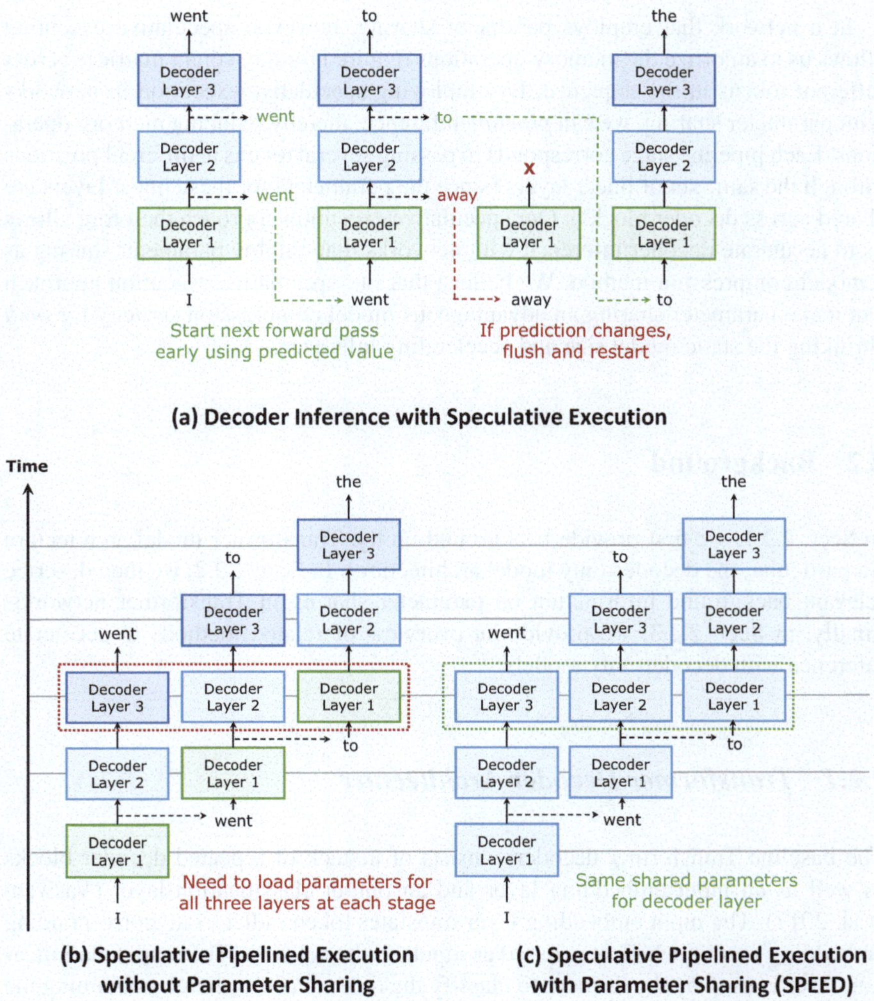

Fig. 2.2 Outline of our method for speculative pipelined execution with parameter sharing. Diagram **a** shows how speculative values can be used to start later tokens, and how any incorrect predictions can later be corrected. Diagram **b** shows how this type of speculative execution allows us to pipeline inference, thereby achieving parallelism across the sequence length dimension. However, in a standard decoder, this doesn't help reduce MOPs since we would now need to load different decoder layers for different tokens in the sequence. Diagram **c** shows how in networks with parameter sharing, speculative pipelined execution amortizes MOPs across the sequence length dimension, thereby allowing for efficient decoding (SPEED)

In a network that employs parameter sharing, however, speculative execution allows us to amortize the memory operations required for the weight matrices across different tokens in the sequence. By employing speculative execution in networks with parameter sharing, we can *pipeline* inference, thereby reducing memory operations. Each pipeline stage corresponds to passing several tokens at different positions through the same set of linear layers (since the parameters for these linear layers are shared across decoder blocks). Our speculative execution approach therefore allows us to accelerate decoder inference with networks that employ parameter sharing as a model compression method. We believe that our speculative execution approach can make parameter sharing an advantageous model compression strategy for *both* shrinking the static model size and accelerating inference.

2.2 Background

In Sect. 2.2.1, we first provide background on the Transformer model architecture (in particular, the decoder-only model architecture). In Sect. 2.2.2, we then describe relevant background information on parameter sharing in Transformer networks. Finally, in Sect. 2.2.3, we provide an overview of related methods to accelerate inference with decoder-only models.

2.2.1 Transformer Decoder Architecture

The baseline Transformer decoder consists of a stack of repeated decoder blocks as well as an input embedding layer and an output classification layer (Vaswani et al. 2017). The input embedding layer translates token indices into corresponding embedding vectors which are passed as inputs to the network. The output classification layer is used by the decoder to classify the output vector (thereby obtaining the next token index). In a decoder-only model, the decoder is used both to process the input sequence in parallel as well as to generate the output sequence one token at a time.

When inferring a model autoregressively, a decoding algorithm is used to determine which token to pass into the model on the next iteration. In the simplest case (referred to as "greedy decoding" (Holtzman et al. 2020), the token is classified as the index of the highest value in the output vector. Another common decoding algorithm is "beam search" (Jurafsky and Martin 2023, Chapter 10), which keeps track of multiple most likely sequences. Additionally, when inferring an autoregressive Transformer decoder, we need to store the previous key and value activations (since these are required for future decoding iterations, and it would be prohibitively expensive to regenerate them at each iteration). These stored keys and values are referred to as the key/value (KV) cache.

2.2.2 Parameter Sharing

Several prior works have explored parameter sharing in Transformer networks as a method for reducing the size of the network. ALBERT (Lan et al. 2020) originally proposed parameter sharing as a method of reducing model size for encoder-only networks. The Universal Transformer (Dehghani et al. 2019) instead applied parameter sharing to both the encoder and decoder (starting from a baseline Transformer architecture). Reid et al. (2021) also applied parameter sharing in the decoder-only context. Dehghani et al. (2019), Lan et al. (2020), and Reid et al. (2021) shared parameters across all layers in the encoder and/or all layers in the decoder. Takase and Kiyono (2023) explored only sharing parameters amongst a subset of layers and found that *cyclic* parameter sharing schemes outperformed sharing across all layers. Cyclic parameter sharing refers to a network with N_d distinct layers which are stacked G times (giving a total number of layers of $N = N_d \times G$).

2.2.3 Accelerating Autoregressive Inference

There have been several works on performing speculative execution in decoder inference (Leviathan et al. 2023, Xia et al. 2023; Kim et al. 2024). These methods aim to produce a set of 'draft' tokens autoregressively using a smaller network (or a subset of a larger network) and then correct them (in parallel) using a larger network. Our work instead aims to support speculative execution within a single network in order to accelerate inference with parameter sharing networks.

There are also prior works that aim to accelerate decoder inference through early exiting (Schuster et al. 2022, Tang et al. 2023). These works aim to accelerate decoder inference by terminating inference early when the model is confident that it can already predict the next token. Our work also leverages similar intuition, namely that while certain predictions truly benefit from the model's full capacity, other continuations are more trivial and can be solved with reduced compute (Schuster et al. 2022). However, our proposed approach for accelerating decoder inference has advantages over typical early exiting approaches. Although early exiting can be applied to an existing network and doesn't require pretraining, our method is guaranteed to always achieve the same accuracy as the baseline network with parameter sharing since it fixes any mistakes. Our method also provides a reduction in model size through parameter sharing in the decoder (in addition to the speedup from pipelined execution).

2.3 Method

This section outlines the methodology for SPEED. Section 2.3.1 outlines how we perform parameter sharing in order to facilitate our speculative inference method. Section 2.3.2 then outlines how we incorporate a novel speculative decoding strategy in order to enable amortizing MOPs along the sequence length dimension.

2.3.1 *Parameter Sharing*

The parameter sharing scheme in this work corresponds to the "CYCLE" configurations from Takase and Kiyono (2023), meaning that if a group of two decoder layers is shared three times, a forward pass consists of alternating between going through layer 1 and layer 2 three times. Our cyclic parameter sharing scheme is outlined graphically in Fig. 2.3a. During fine-tuning, we incorporate a weighted loss function inspired by the work of Schuster et al. (2022). The purpose is to adapt the output classifier so that it can make early predictions during inference using the output logits from earlier decoder layers (i.e., after different repetitions of decoder layer groups). More formally, the shared loss function is given by $L_w = \sum_{i=1}^{G} w_i L_i$, where G is the number of decoder layer groups, and L_i and w_i correspond to the loss and the applied

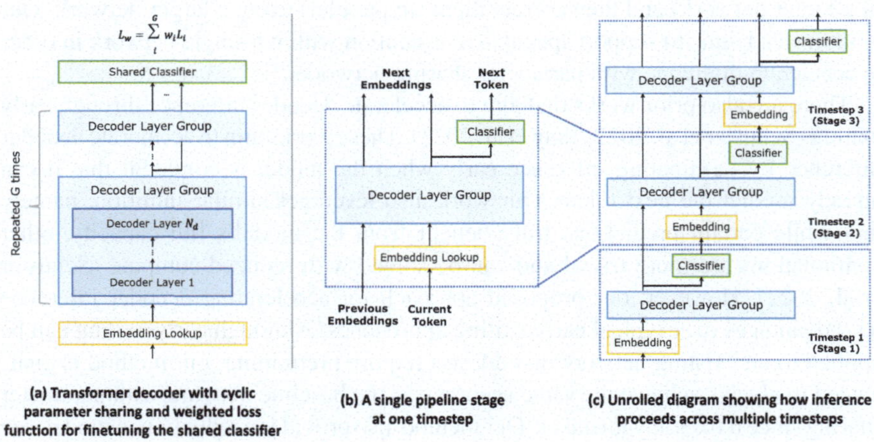

Fig. 2.3 A demonstration of how our speculative pipelined execution approach is implemented. Diagram **a** shows how parameters are shared cyclicly across groups of decoder layers (where N_d is the number of unique decoder layers shared G times and $N = G * N_d$ is the total number of layers), and how the losses from the classifications at each layer are incorporated into a shared loss function during training. Note that we have omitted the final layer normalization layer prior to the shared classifier for simplicity. Diagram **b** shows a single pipeline stage in our inference process. Note that all embeddings must pass through the classifier for the invalidation logic (Sect. 2.3.2), although we have omitted them from the illustration for simplicity. Diagram **c** shows the progression of several pipeline stages in sequence

weighting for group i, respectively. The default weighting scheme we use is the linear weighting described in Schuster et al. (2022), which is given as $w_i = i/(\sum_i i)$. Note that this weighting scheme intentionally weights the loss for later layers higher to ensure the final output accuracy is not degraded. As T5 uses pre-norm and has a final layernorm after the last decoder layer, we applied the final normalization layer to the output embeddings from each layer before passing them into the shared classifier. The training scheme using a shared classifier is also illustrated in Fig. 2.3a.

2.3.2 Speculative Pipelined Execution

Figure 2.3b and c outline how the forward pass is performed in our speculative approach. In essence, SPEED speculatively predicts future tokens based on early predictions and then concatenates them with the current token for their parallel processing. The crucial feature of SPEED is its *invalidation logic* since speculative predictions can sometimes be incorrect. To achieve this, our framework keeps track of previous classifications for each token at the previous stage (i.e., before passing through a decoder layer group) and performs the invalidation logic whenever subsequent classifications change after the current stage (i.e., after passing through a decoder layer group). In such a case, any future iterations that have been speculatively initiated using the previous classifications must be flushed out and restarted.

Another crucial implementation detail is that the internal logic in the attention module and the internal KV cache management logic both need to be modified to facilitate pipelining. The KV cache corresponds to intermediate activations associated with earlier tokens in the sequence, which are required for calculating later tokens. The KV cache management logic has to ensure that when future tokens are invalidated, all previous KV cache updates corresponding to these tokens are also invalidated. These modifications play a key role in ensuring that the final output classification for each token remains unaffected by speculation.

2.4 Experimental Setup

Sections 2.4.1 and 2.4.2 provide relevant implementation details for our experiments.

2.4.1 Training Details

We used the baseline T5-Base decoder-only model architecture (Raffel et al. 2020), which has 12 decoder layers, a hidden dimension of 768, 12 attention heads each with dimension 64, and an FFN dimension of 2048 (the default configuration in T5X (Roberts et al. 2023). We used the default SentencePiece tokenizer with a vocabulary size of 32K, and we used tied input and output embeddings (Kudo and Richardson 2018). We pretrained the model on the C4 dataset for 524,288 steps using a batch size of 128 (Raffel et al. 2020). We used a base learning rate of 1 with a square root decaying learning rate scheduler and with 10K warmup steps.

For downstream tasks, we focused on two particular sequence-to-sequence tasks during finetuning: translation and summarization. For translation, we used the WMT English to German dataset as well as the English to German Paracrawl-Paragraph translation dataset (Barrault et al. 2019, Ghussin et al. 2023). The Paracrawl-Paragraph dataset was used to evaluate a translation dataset with longer source and target context lengths as it consists of full-paragraph translations. For summarization, we used the CNN/DailyMail dataset (Hermann et al. 2015), as well as the English-to-English split of the Wikilingua multilingual summarization dataset (Ladhak et al. 2020). We finetuned for 262,144 steps using a batch size of 128 and dropout of 0.1, using input/target sequence lengths of 512/512 across all tasks. When finetuning, we used a constant learning rate of 0.001 with 1K warmup steps. We evaluated checkpoints every 5,000 steps during finetuning on the validation set, and then reported results on the test set using the checkpoint with the best accuracy on the validation set. Both training and inference arithmetic were performed in BF16 precision. We used TPU v2-8 machines on Google Cloud Platform for training experiments, which are launched using Skypilot (Yang et al. 2023).

2.4.2 *Implementation Details*

We implement SPEED within the T5X (Roberts et al. 2023) repository, which is built on top of the JAX (Bradbury et al. 2018) framework. Our implementation for pipelined inference used a custom decoding function in the T5X framework. Our initial profiling runs also indicated that the existing greedy decoding function in JAX had greater runtime overhead than our custom decoding algorithm, likely due to additional optional arguments that were unused in our experiments. In order to benchmark the networks without parameter sharing, we therefore implemented a stripped-down greedy decoding function to serve as a fair baseline since it has minimal added control logic.

2.5 Results

Section 2.5.1 provides the main results from our work, showcasing the benefits of our method in terms of accuracy and efficiency, as well as additional discussion of performance on particular datasets. Section 2.5.2 provides detailed analysis of the prediction accuracy at different layers in the network. Finally, Sect. 2.5.3 discusses the factors that affect the latency benefits when employing SPEED.

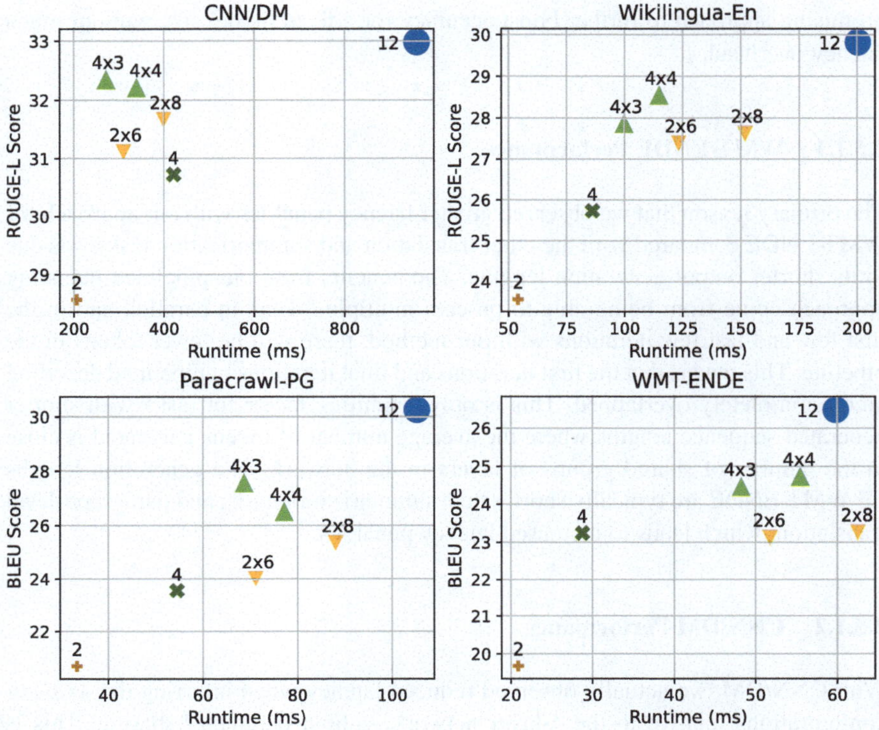

Fig. 2.4 Accuracy versus Efficiency for T5-Base with Speculative Execution on an NVIDIA A5000 GPU. The stars correspond to the parameter sharing configurations with speculative pipelined execution, and the circles correspond to the baseline configurations (with either the normal full-length decoder or a shallower decoder). The dot sizes are proportional to the number of parameters in the decoder. The arrows indicate the accuracy improvements we can get from using parameter sharing without the full latency penalty that we would normally incur. The reported runtime is the average latency across 500 examples in the test set

2.5.1 Main Results

Figure 2.4 shows the accuracy versus efficiency tradeoff comparisons. When employing parameter sharing with SPEED, we observe significant speedups relative to the baseline 12-layer decoder network of up to $\sim 3\times$, achieving close to the same runtime as the shorter decoder network without parameter sharing across all benchmarks other than WMT-ENDE. Additionally, our parameter-sharing configurations attain significantly higher accuracy than the shallow decoder baselines. This demonstrates how the SPEED approach allows for improving accuracy for a fixed model size without a significant runtime penalty. We further experiment with deepening the decoder with parameter sharing by sharing parameters more times such that the total number of layers is increased. We find that deepening the decoder generally improves accuracy with minimal runtime penalty; as such, we believe that this is a promising approach to further boost accuracy for a fixed model size without much latency overhead.

2.5.1.1 WMT-ENDE Performance

The primary reason that we observed greater latency penalties with our approach for WMT-ENDE compared with the other translation and summarization tasks was due to its shorter output generation lengths. The benefits from our pipelined decoding approach come from being able to process multiple tokens in parallel, and in the first few and last few iterations with our method, there will be fewer tokens in the pipeline. This means that the first iterations and final iterations in pipelined decoding aren't completely overlapped. This is only a limiting factor for tasks with shorter generated sequence lengths where the average number of tokens generated is close to the number of shared groups of layers in the network. The generation lengths for WMT-ENDE are typically shorter than summarization tasks and paragraph-level translation, which leads to increased latency penalties.

2.5.1.2 CNN/DM Performance

With CNN/DM, we actually observed reduced latency when inferring the 4x3/4x4 configurations relative to the 4-layer network without parameter sharing. This is unexpected, since even assuming perfect prediction for the networks with parameter sharing, the latency would not be less than the baseline 4-layer network assuming the same output generation length. However, it is possible for the parameter sharing configurations to exhibit lower latency due to differences in the average generation lengths for the networks with parameter sharing relative to the network without parameter sharing.

2.5.2 *Prediction Consistency*

In order to assess the accuracy of the predictions made from our network at earlier layers, we profiled the proportion of predictions that were flipped between pairs of layers during inference. Figure 2.5 shows the prediction consistencies for 4x3 and 2x6 network configurations across all tasks. Upon examining these numbers and plots, we found that the model is able to make the majority of predictions accurately at early layers. Across all three configurations, the proportion of predictions that would need to be corrected after the first layer was between 13–17% for the 2x6 configuration and between 6–14% for the 4x3 configuration, showing that the majority of predictions were correct at early layers. Additionally, we found that a very small percentage of predictions flipped at later layers. This shows that the model tends to converge to the final answer and does not experience much oscillation between different predictions.

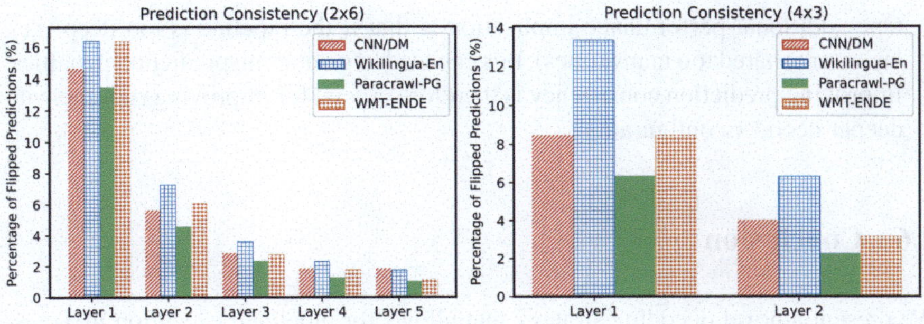

Fig. 2.5 Proportion of predictions that were flipped at each layer when using speculative pipelined execution with 2x6 and 4x3 configurations

Y_{pred}: Sally Forrest, whose birth name was Katherine Feeney, was 86 and had long battled cancer. She appeared in as herself in an episode of The Ed Sullivan Show and three episodes of The Dinah Shore Chevy Show . Other , her IMDb page says. She also appeared in a Broadway production of The Seven Year Itch.

Fig. 2.6 Visualizing the model's early incorrect predictions that are later corrected, using the 4x3 decoder-only model evaluated on a sample from the CNN/DM dataset. The green highlighted words are predictions that were correct from the start (i.e., after the first layer group), while the red highlighted (underlined) words are predictions that were wrong and had to be corrected later in the network. The majority of predictions are correct after the first layer group

Figure 2.6 provides a qualitative example of the model's prediction consistency, demonstrating that the majority of predictions can be made accurately at early layers.

2.5.3 General Discussion

There are several factors that impact the runtime when employing SPEED.

- One factor is the generation length, as the first iterations and final iterations in pipelined decoding aren't completely overlapped (as discussed in Sect. 2.5.1). This limits the runtime gains from SPEED for tasks with short output generation lengths.
- Because the embedding matrix is large, it can actually end up consuming a large portion of the memory bandwidth (and hence the runtime) for smaller models. This is a crucial reason why the latency gains aren't linear as you go from a 12-layer network down to a 2-layer network even without considering parameter sharing or speculative execution.

- One additional performance implication is that if the pipeline is too deep (i.e., layers are shared too many times), this can lead to greater misprediction penalties. Improving prediction consistency is therefore crucial for improving runtime with deeper decoder configurations.

2.6 Conclusion

We present a novel decoding strategy that allows for pipelined execution in Transformer decoders with parameter sharing. We describe the modifications to the model architecture that are required to leverage pipelined execution in order to reduce memory traffic (namely, cyclic parameter sharing in the decoder module). We observe consistent accuracy gains across all tasks for an equivalent model size, with only a small latency penalty. These results demonstrate the accuracy and performance benefits of our pipelined inference approach, showing how SPEED allows for deeper decoder configurations with parameter sharing to improve accuracy for a fixed parameter budget and minimal latency penalty.

Acknowledgements We acknowledge gracious support from Intel, Furiosa, Apple, and Samsung SAIT. We also appreciate the support from Microsoft through their Accelerating Foundation Model Research, including great support from Sean Kuno. Furthermore, we appreciate support from Google Cloud, the Google TRC team, and specifically Jonathan Caton, and Prof. David Patterson. Prof. Keutzer's lab is sponsored by the Intel corporation, Intel One-API, Intel VLAB team, the Intel One-API center of excellence, as well as funding through BDD and BAIR. Sehoon Kim would like to acknowledge the support from the Korea Foundation for Advanced Studies (KFAS). Amir Gholami was supported through funding from Samsung SAIT. Our conclusions do not necessarily reflect the position or the policy of our sponsors, and no official endorsement should be inferred.

References

Al Ghussin Y, Zhang J, van Genabith J (2023) Exploring paracrawl for document-level neural machine translation. In: Proceedings of the 17th conference of the european chapter of the association for computational linguistics. Association for Computational Linguistics, Dubrovnik, Croatia, pp 1304–1310. https://aclanthology.org/2023.eacl-main.94

Barrault L, Bojar O, Costa-jussà MR, Federmann C, Fishel M, Graham Y, Haddow B, Huck M, Koehn P, Malmasi S, Monz C, Müller M, Pal S, Post M, Zampieri M (2019) Findings of the 2019 conference on machine translation (WMT19). In: Proceedings of the fourth conference on machine translation (Volume 2: Shared Task Papers, Day 1), Florence, Italy, pp 1–61. http://www.statmt.org/wmt19/translation-task.html

Bradbury J, Frostig R, Hawkins P, Johnson MJ, Leary C, Maclaurin D, Necula G, Paszke A, VanderPlas J, Wanderman-Milne S, Zhang Q (2018) JAX: composable transformations of Python+NumPy programs. http://github.com/google/jax

Brown T, Mann B, Ryder N, Subbiah M, Kaplan JD, Dhariwal P, Neelakantan A, Shyam P, Sastry G, Askell A, Agarwal S, Herbert-Voss A, Krueger G, Henighan T, Child R, Ramesh A, Ziegler D, Wu J, Winter C, Hesse C, Chen M, Sigler E, Litwin M, Gray S, Chess B, Clark J, Berner C, McCandlish S, Radford A, Sutskever I, Amodei D (2020) Language models are few-shot learners. In: Advances in neural information processing systems, vol 33, pp 1877–1901. https://proceedings.neurips.cc/paper_files/paper/2020/file/1457c0d6bfcb4967418bfb8ac142f64a-Paper.pdf

Dehghani M, Gouws S, Vinyals O, Uszkoreit J, Kaiser L (2019) Universal transformers. In: ICLR (Poster), OpenReview.net. http://dblp.uni-trier.de/db/conf/iclr/iclr2019.html#DehghaniGVUK19

Devlin J, Chang MW, Lee K, Toutanova K (2019) BERT: Pre-training of deep bidirectional transformers for language understanding. In: Proceedings of the 2019 conference of the North American chapter of the association for computational linguistics (ACL): human language technologies, Vol 1 (Long and Short Papers), pp 4171–4186

Hermann KM, Kocisky T, Grefenstette E, Espeholt L, Kay W, Suleyman M, Blunsom P (2015) Teaching machines to read and comprehend. Advances in neural information processing systems 28

Holtzman A, Buys J, Du L, Forbes M, Choi Y (2020) The curious case of neural text degeneration. In: International conference on learning representations (ICLR). https://openreview.net/forum?id=rygGQyrFvH

Jurafsky D, Martin JH (2023) Speech and language processing, 3rd edn. Prentice Hall. https://web.stanford.edu/~jurafsky/slp3/

Kim S, Hooper C, Wattanawong T, Kang M, Yan R, Genç H, Dinh G, Huang Q, Keutzer K, Mahoney MW, Shao YS, Gholami A (2023) Full stack optimization of transformer inference: a survey. arXiv:abs/2302.14017. https://api.semanticscholar.org/CorpusID:257219934

Kim S, Mangalam K, Moon S, Malik J, Mahoney MW, Gholami A, Keutzer K (2024) Speculative decoding with big little decoder. Advances in neural information processing systems 36

Kudo T, Richardson J (2018) SentencePiece: a simple and language independent subword tokenizer and detokenizer for neural text processing. In: Proceedings of the 2018 conference on empirical methods in natural language processing: system demonstrations. Association for Computational Linguistics, Brussels, Belgium, pp 66–71. https://aclanthology.org/D18-2012

Ladhak F, Durmus E, Cardie C, McKeown K (2020) WikiLingua: a new benchmark dataset for cross-lingual abstractive summarization. arXiv preprint arXiv:2010.03093

Lan Z, Chen M, Goodman S, Gimpel K, Sharma P, Soricut R (2020) ALBERT: a lite BERT for self-supervised learning of language representations. In: 8th international conference on learning representations, ICLR 2020, Addis Ababa, Ethiopia, April 26–30, 2020, OpenReview.net. https://openreview.net/forum?id=H1eA7AEtvS

Leviathan Y, Kalman M, Matias Y (2023) Fast inference from transformers via speculative decoding. In: Proceedings of the 40th international conference on machine learning, JMLR.org

Park J, Yoon H, Ahn D, Choi J, Kim JJ (2020) OPTIMUS: OPTImized matrix MUltiplication Structure for Transformer neural network accelerator. Proc Mach Learn Syst 2:363–378

Radford A, Wu J, Child R, Luan D, Amodei D, Sutskever I, et al. (2019) Language models are unsupervised multitask learners, p 9

Raffel C, Shazeer N, Roberts A, Lee K, Narang S, Matena M, Zhou Y, Li W, Liu PJ (2020) Exploring the limits of transfer learning with a unified text-to-text transformer. J Mach Learn Res 5485–5551

Reid M, Marrese-Taylor E, Matsuo Y (2021) Subformer: exploring weight sharing for parameter efficiency in generative transformers. In: Findings of the association for computational linguistics: EMNLP, Punta Cana, Dominican Republic, pp 4081–4090. https://aclanthology.org/2021.findings-emnlp.344

Roberts A, Chung HW, Mishra G, Levskaya A, Bradbury J, Andor D, Narang S, Lester B, Gaffney C, Mohiuddin A et al (2023) Scaling up models and data with t5x and seqio. J Mach Learn Res 24(377):1–8

Schuster T, Fisch A, Gupta J, Dehghani M, Bahri D, Tran V, Tay Y, Metzler D (2022) Confident adaptive language modeling. Adv Neural Inf Process Syst 35:17456–17472

Takase S, Kiyono S (2023) Lessons on parameter sharing across layers in transformers. In: Proceedings of the fourth workshop on simple and efficient natural language processing (SustaiNLP), Toronto, Canada (Hybrid), pp 78–90. https://aclanthology.org/2023.sustainlp-1.5

Tang S, Wang Y, Kong Z, Zhang T, Li Y, Ding C, Wang Y, Liang Y, Xu D (2023) You need multiple exiting: dynamic early exiting for accelerating unified vision language model. In: Proceedings of the IEEE/CVF conference on computer vision and pattern recognition, pp 10781–10791

Vaswani A, Shazeer NM, Parmar N, Uszkoreit J, Jones L, Gomez AN, Kaiser L, Polosukhin I (2017) Attention is all you need. In: Neural information processing systems. https://api.semanticscholar.org/CorpusID:13756489

Wolf T, Debut L, Sanh V, Chaumond J, Delangue C, Moi A, Cistac P, Rault T, Louf R, Funtowicz M, Davison J, Shleifer S, von Platen P, Ma C, Jernite Y, Plu J, Xu C, Le Scao T, Gugger S, Drame M, Lhoest Q, Rush A (2020) Transformers: state-of-the-art natural language processing. In: Proceedings of the 2020 conference on empirical methods in natural language processing: system demonstrations, Online, pp 38–45. https://aclanthology.org/2020.emnlp-demos.6

Xia H, Ge T, Wang P, Chen SQ, Wei F, Sui Z (2023) Speculative Decoding: Exploiting Speculative Execution for Accelerating Seq2seq Generation. In: Findings of the association for computational linguistics: EMNLP 2023, Singapore, pp 3909–3925. https://aclanthology.org/2023.findings-emnlp.257

Yang Z, Wu Z, Luo M, Chiang WL, Bhardwaj R, Kwon W, Zhuang S, Luan FS, Mittal G, Shenker S, Stoica I (2023) SkyPilot: an intercloud broker for sky computing. In: 20th USENIX symposium on networked systems design and implementation (NSDI 23), Boston, MA, pp 437–455. https://www.usenix.org/conference/nsdi23/presentation/yang-zongheng

Chapter 3
Efficient LLM Inference on CPUs

Haihao Shen, Hanwen Chang, Bo Dong, Yu Luo, and Hengyu Meng

Abstract LLMs have demonstrated remarkable performance with tremendous potential across a wide range of tasks. However, deploying these models has been challenging due to the astronomical number of model parameters, which necessitates large memory capacity and high memory bandwidth. In this chapter, we propose an effective approach to make the deployment of LLMs more efficient. We support automatic INT4 weight-only quantization and design a specialized LLM Runtime with highly optimized kernels to accelerate LLM inference on CPUs. We demonstrate the general applicability of our approach on popular LLMs, including Llama (Touvron et al. 2023a), Llama2 (Touvron et al. 2023b), and Mistral (Jiang et al. 2023), showcasing extreme inference efficiency on CPUs. The code is publicly available at: https://github.com/intel/intel-extension-for-transformers.

3.1 Introduction

LLMs have shown remarkable performance with tremendous potential across a wide range of tasks (Roziere et al. 2023). However, deploying these models has been challenging due to the astronomical number of model parameters, which necessitates significant memory capacity and high memory bandwidth.

Quantization is a technique used to reduce the numeric precision of weights and activations of a neural network to lower the computation costs of inference. INT8 quantization (Vanhoucke et al. 2011) is the most widely used approach today, given the trade-off between high inference performance and reasonable model accuracy. However, outliers in activations limit the general adoption of INT8 quantization, though some related works have addressed these issues (Xiao et al. 2023). FP8, a newly introduced data type, has attracted much attention (Micikevicius et al. 2022) but has seen limited adoption due to hardware unavailability. Conversely, weight-only quantization has become popular as it applies low precision (e.g., 4-bit) to

---

H. Shen · H. Chang (✉) · B. Dong · Y. Luo · H. Meng
Intel Corporation, No. 880 Zi Xing Road, 200241 Shanghai, China
e-mail: hanwen.chang@intel.com

© The Author(s), under exclusive license to Springer Nature Switzerland AG 2025
P. Passban et al. (eds.), *Enhancing LLM Performance*, Machine Translation: Technologies and Applications 7, https://doi.org/10.1007/978-3-031-85747-8_3

weights only, while maintaining higher precision (e.g., 16-bit floating point) for activations, thus preserving model accuracy. Many excellent works on 4-bit weight-only quantization (Dettmers et al. 2023) have demonstrated effectiveness in LLM inference. Meanwhile, the open-source community is embracing such low-bit weight-only quantization and has made available CPP-based implementations such as llama.cpp[1] and starcoder[2] based on the GGML library. These implementations, typically optimized for CUDA, may not work efficiently on CPUs. Therefore, it is crucial to address the challenge of making LLM inference efficient on CPUs.

In this research, we propose an effective approach for LLM inference on CPUs including an Automatic INT4 Quantization Process and an efficient LLM Runtime. We leverage the Intel neural compressor[3] that provides the support of INT4 quantization such as GPTQ (Frantar et al. 2022), AWQ (Lin et al. 2023), TEQ (Cheng et al. 2023a), SignRound (Cheng et al. 2023b) and generate the INT4 model automatically. Inspired by the GGML[4] library, we develop a tensor library for CPU, supporting all the mainstream instruction sets such as AVX2, AVX512, AVX512_VNNI (Rodriguez et al. 2018), and advanced matrix extensions.[5] Our results show that the average latency for generating tokens ranges from 20ms to 80ms on LLMs with 6B to 20B parameters, using just a single socket of fourth generation Intel Xeon scalable processors. while preserving high accuracy with only a 1% loss from the FP32 baseline. Our main contributions are as follows:

- We propose an automatic INT4 quantization flow and generate high-quality INT4 models with negligible accuracy loss within 1% from the FP32 baseline.
- We design a tensor library that supports general CPU instruction sets as well as the latest instruction sets for deep learning acceleration. With a new CPU tensor library, we develop an efficient LLM Runtime model to accelerate inference.
- We apply our inference solution to popular LLM models covering 3B to 20B models and demonstrate the promising per-token generation latency from 20ms to 80ms.

The remainder of this chapter is organized as follows. Section 3.2 describes the limitations of prior works and discusses how we address them. Section 3.3 introduces the approach, which includes INT4 quantization and inference. Section 3.4 outlines the experimental setup, presents accuracy and performance results, and discusses performance tuning. Section 3.5 presents the conclusions and outlines future work.

---

[1] https://github.com/ggerganov/llama.cpp.

[2] https://github.com/bigcode-project/starcoder.cpp.

[3] https://github.com/intel/neural-compressor.

[4] Group-wise gradient-based mix-bit low-rank: https://github.com/ggerganov/ggml.

[5] https://www.intel.com/content/www/us/en/products/docs/accelerator-engines/advanced-matrix-extensions/overview.html.

3.2 Related Work

GGML is a quantization technique that assigns different bit-widths to various weight groups to significantly reduce memory usage. In LLM inference on CPU, the most significant cost is often I/O bound, so reducing memory usage can greatly improve performance. llama.cpp is a GGML-based implementation that accelerates low-bit inference. This implementation is very popular because CPUs can use it to perform inference on LLMs. The GGML-based implementation has limitations, including accuracy, performance, and extensibility.

Although GGML provides different bit-widths for different groups, the accuracy has not been validated across various tasks. Using 4 bits with GGML, many LLMs still cannot achieve less than a 1% accuracy loss. We validated numerous LLMs on various tasks and calculated the average to compare INT4 and FP32 models. GGML-INT4 often struggles to achieve good accuracy with models like Llama-2-7b-hf and GPTNeoX-20b. Using 3 bits or other combinations may result in even worse accuracy. GPTQ (Frantar et al. 2022), AWQ (Lin et al. 2023) and other advanced quantization methods can improve accuracy; however, the GGML-based implementation cannot automatically generate them; users need to convert it themselves. Even with advanced quantization, we found that some LLMs still cannot achieve satisfactory accuracy.

As far as performance is concerned, while prior work may have outperformed FP16 or BF16, upon further investigation into the implementation, we believe CPUs can achieve even higher performance. GGML-based implementation doesn't utilize vector instructions such as VNNI and AMX. The default quantization group size is set to 32 for better accuracy. However, using a group size of 128 or per-channel quantization can significantly improve performance if the accuracy issue can be resolved. Additionally, the implementation lacks fusion kernels like MHA, which is why the first token latency is still poor. Finally, the GGML-based implementation did not have the ability to extend the solution to 2-sockets on CPU or multi-nodes.

For accuracy, we introduced an automatic INT4 quantization flow to address accuracy issues and improve usability. For performance, we developed an efficient LLM called Runtime to enhance both first-token and next-token latency. For extensibility, we implemented tensor parallelism for inference across 2 sockets using 4-bit models.

3.3 Method

In this section, we introduce the approach, which consists of two major components: an automatic INT4 quantization flow and an efficient LLM Runtime, as shown in Fig. 3.1. More details are provided in the subsequent sections.

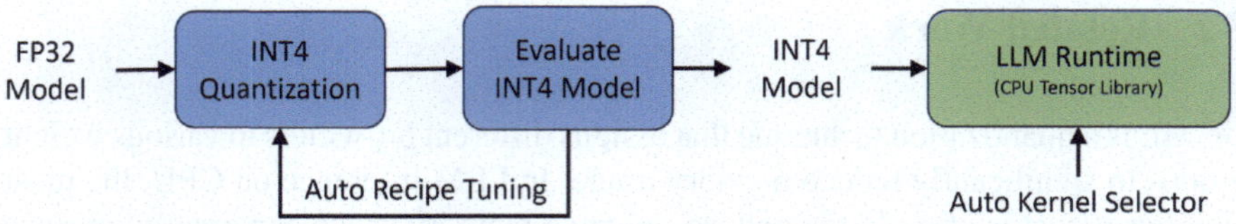

Fig. 3.1 The left part depicts the automatic INT4 quantization flow. Starting with an FP32 model, this flow utilizes the default INT4 quantization recipes to evaluate the INT4 model's accuracy; the recipe tuning loop is optional if the INT4 model meets the accuracy target. The right part shows a simplified runtime for efficient LLM inference, built on top of a CPU tensor library that includes an automatic kernel selector

3.3.1 Automatic INT4 Quantization Flow

The INT4 quantization flow was developed with the Intel neural compressor, a popular quantization tool for deep learning frameworks. Since the tool supports mainstream INT4 quantization methods such as GPTQ (Frantar et al. 2022), SignRound (Cheng et al. 2023b), AWQ (Lin et al. 2023), TEQ (Cheng et al. 2023a), and RTN (round-to-nearest), our automatic quantization flow allows tuning of various quantization methods and granularities (channel-wise or group-wise), different group sizes (32, 64, 128 ... 1024) as well as which linear operators should not be quantized. Each method generates an INT4 model, which is then evaluated. Once the INT4 model meets the accuracy target, it is passed to LLM Runtime for performance evaluation. We evaluated approximately 100 LLMs, and most of them achieved a low accuracy loss, typically less than one percent. Additionally, we propose a state-of-the-art INT4 algorithm called AutoRound, which has demonstrated superior performance compared to other algorithms. We experimented with 3-bit quantization but found it more challenging to achieve acceptable accuracy compared to 4-bit. While using a smaller group size can produce a good INT4 model, this is often at the expense of performance. When considering the balance between accuracy and performance, we find that INT4 offers a compelling LLM solution.

3.3.2 Efficient LLM Runtime

LLM Runtime is designed to enable efficient inference of LLMs on CPUs. Figure 3.2 illustrates the key components of LLM Runtime, where the components (CPU tensor library and LLM optimizations) in green (labeled with •) are specialized for LLM inference, while those in blue (without •) are essential for general runtime functionality. More details about the CPU tensor library are described in the following paragraphs. Note that the design is flexibly extensible, with a hardware abstraction layer (currently for CPU only), while the support for other hardware types is beyond the scope of this study.

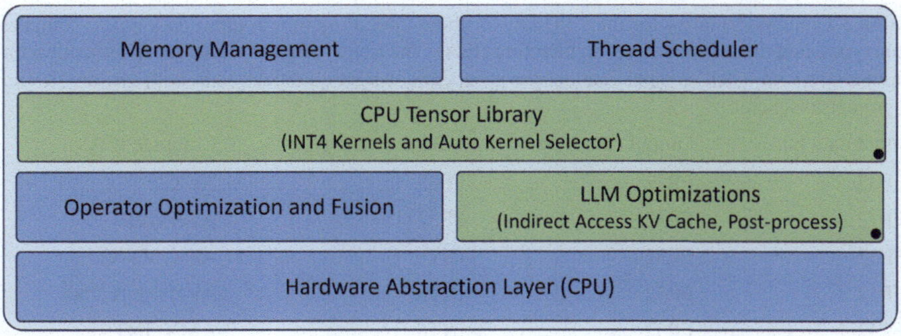

Fig. 3.2 Key components in LLM Runtime including general components and LLM specialized components (labeled with •)

3.3.3 CPU tensor library in LLM Runtime

We developed a CPU tensor library for linear algebra subroutines, inspired by the template design of Cutlass.[6] The tensor library is built using Xbyak, a just-in-time (JIT) assembler for x86 architectures. As shown in Table 3.1, this library provides extensive support for INT4 kernels on x86 CPUs, with AMX available in the latest Intel Xeon scalable processors and VNNI available in both Intel and AMD CPUs. The CPU tensor library offers more than just GEMM operations; it also includes a prologue and an epilogue. With these components, the library can replace the matrix multiplication operation with an INT4 operation. The prologue dequantizes the weight matrix from low-bit format to the computation data type and can also quantize the activation matrix when using AMX INT8 or VNNI. The flow of weight data type INT4 and computation data type INT8 is shown in Fig. 3.3. This library supports kernels for popular fusion patterns such as feedforward networks (FFN), multi-head attention (MHA), and general query attention.

3.3.4 LLM Optimizations

Most recent LLMs typically consist of decoder-only Transformer-based models (Vaswani et al. 2017). Given the unique characteristics of next token generation, the Key/Value (KV) cache becomes performance critical for LLM inference. The optimizations are described in Fig. 3.4. To prevent out-of-memory errors and minimize the costs of switching optimization profiles and changing shapes, LLM Runtime pre-computes the memory requirements for activation tensors at build time. LLM Runtime also integrates continuous batching to enhance throughput. We submitted the results to MLPerf v4.0 GPT-J closed division, highlighting the capabilities of CPUs.

---

[6] https://github.com/NVIDIA/cutlass.

Table 3.1 Support matrix for the CPU tensor library, detailing input/output data types, compute data types, and the instruction set architecture (ISA). The library supports dynamic quantization for input along with batch or input channel per group, and weight quantization in both symmetric and asymmetric scheme

Input data type	Output data type	Compute data type	Compute ISA
FP32	FP32	FP32	AVX2
FP32	FP32	FP32	AVX512F
FP32	FP32	INT8	AVX_VNNI
FP32	FP32	INT8	AVX512_VNNI
FP32	FP32	INT8	AMX_INT8
FP32/FP16	FP32/FP16	FP16	AVX512_FP16
FP32/BF16	FP32/BF16	BF16	AMX_BF16

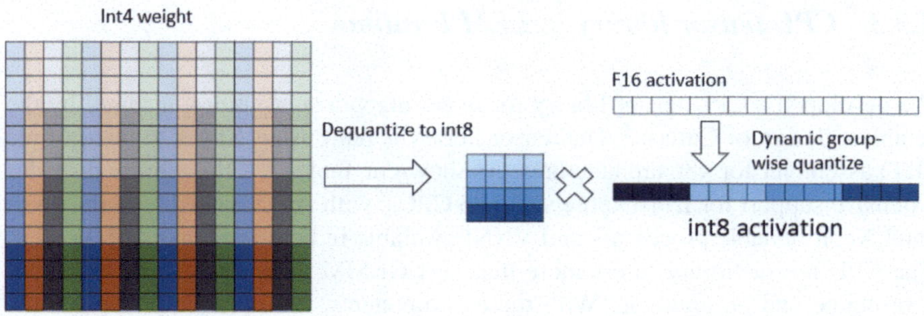

Fig. 3.3 Computation flow of the INT4 op in the CPU tensor library, including *dequant*, *multiplication* and *add*

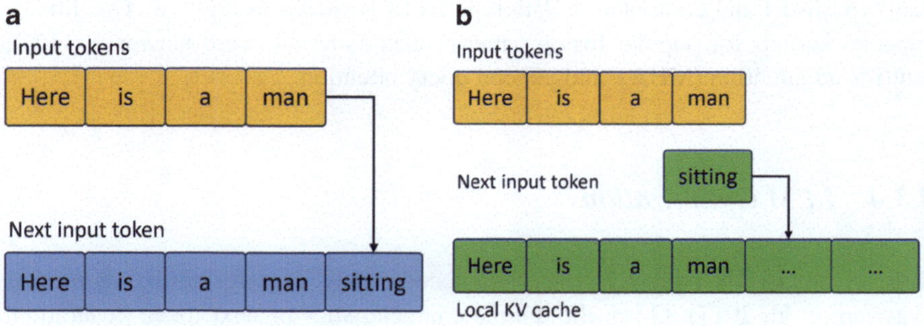

Fig. 3.4 KV cache optimization. Left **a** shows the default KV cache, where new token generation requires memory reallocation for all the tokens (5 in this example); Right **b** shows the optimized KV cache with the pre-allocated KV memory and only new token updated each time

3.3.5 *Tensor Parallelism*

Tensor parallelism is a strategy used to train and perform inference on LLMs by distributing computations/tensors across multiple compute devices. It is a crucial technique for scaling up the size of deep learning models. The CPU tensor library supports tensor parallelism on multiple sockets and nodes, allowing for efficient utilization of computational resources.

When employing tensor parallelism to partition and compute LLMs, various partitioning algorithms can be used. We employ 1D algorithms, which divide matrices by rows or columns. Splitting matrices in row-major order by columns requires data rearrangement, potentially impacting performance. Conversely, splitting such matrices by rows does not incur this overhead.

In our implementation, we pre-split the corresponding weights, so the time spent on this step is one-time and does not affect inference performance. Another significant factor affecting performance is the all reduce operation. Since each node computes partial and incomplete results, an all-reduce operation is necessary to combine the output data. However, this process can be time-consuming. To mitigate this, it's advisable to perform tensor parallelism only on heavy computing operators. This strategy is applied only to FFNs and attention operations. Figure 3.5 illustrates the steps of the FFN process, where we horizontally split the first matrix and vertically split the second matrix, and then combine the results.

Figure 3.6 illustrates the steps of the Attention process, which is more complex. We horizontally split the Q, K, and V matrices, transpose the K Cache, and finally combine the results.

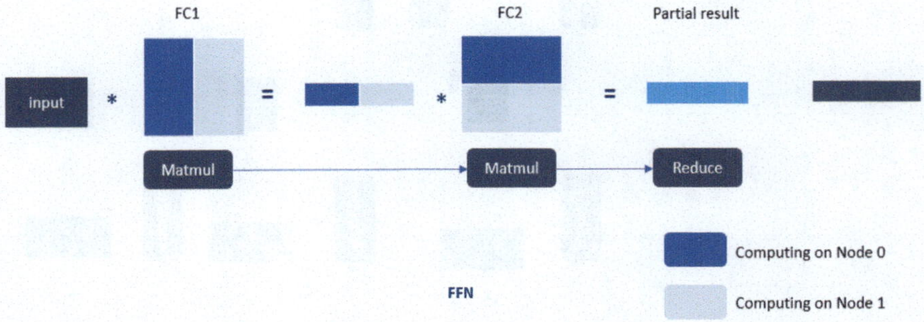

Fig. 3.5 FFN fusion with tensor parallelism, matrix splitting and computation on 2 nodes

Fig. 3.6 Attention with tensor parallelism, matrix splitting and computation on 2 nodes

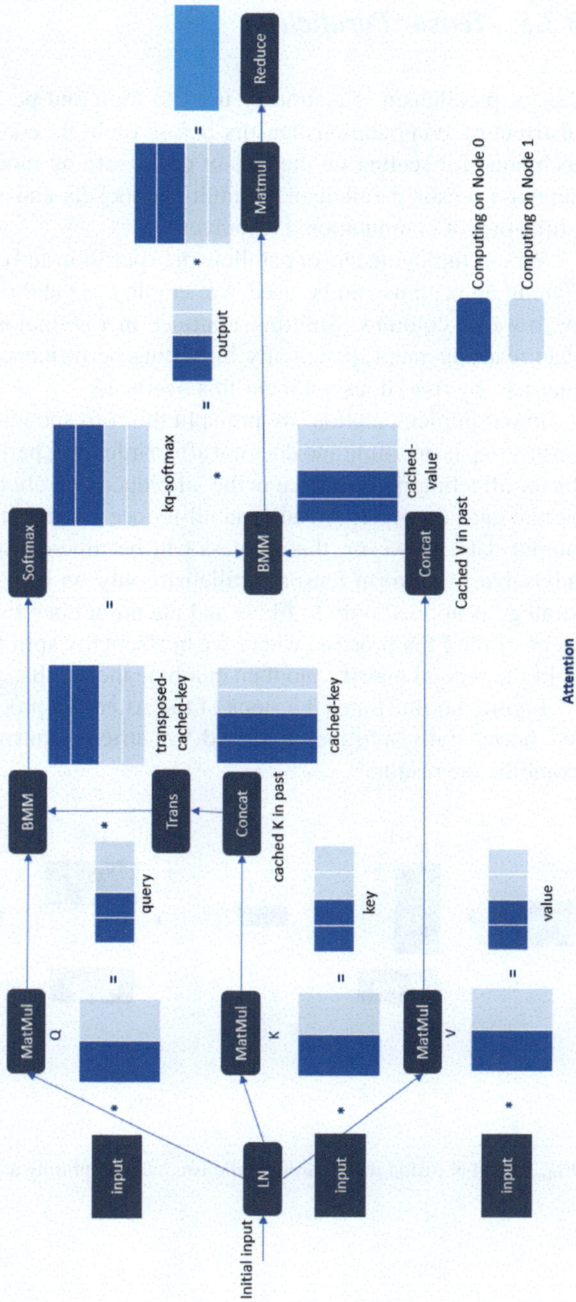

3.4 Results

3.4.1 Experimental Setup

To demonstrate generality, we selected the popular LLMs from a wide range of architectures with model parameter sizes from 7B to 20B. We evaluate the accuracy of both FP32 and INT4 models using open-source datasets from lm-evaluation-harness[7] including Lambada (Paperno et al. 2016) OpenAI, HellaSwag (Zellers et al. 2019), Winogrande (Sakaguchi et al. 2021), Piqa (Bisk et al. 2020), and wikitext.[8] The accuracy data represents the average results of these tasks. We select the best accuracy from various algorithms such as RTN, GPTQ, AWQ, AutoRound, etc. To demonstrate performance, we measure the latency of next token generation on the fourth generation of Intel Xeon® scalable Processors, available on public clouds such as AWS. We conducted experiments with quantization group sizes of 128 and 32. A group size of 32 typically yields better runtime performance but accuracy is lower.

3.4.2 Accuracy

We evaluate the accuracies on the aforementioned datasets and show the average accuracy in Table 3.2. From the table, we observed that the INT4 model achieves accuracies nearly equivalent to the FP32 model, with a relative loss of less than 1% compared to the FP32 baseline. Upon even closer examination, the results remain impressive. For example, GPT-J-6B achieved accuracies of 0.6786 on Lambada-OpenAI, 0.6614 on HellaSwag, 0.648 on Winogrande, and 0.7465 on PIQA. This demonstrates that weight-only INT4 quantization is an effective approach for maintaining accuracy. We validated additional models beyond those listed in the table and consistently observed a loss of less than one percent. Interested readers should refer to our blog post[9] for additional findings.

3.4.3 Memory Usage

Memory usage is a critical metric for the 4-bits solution. We significantly reduce memory usage with quantization and provide the details in Table 3.3. Using a group

---

[7] https://github.com/EleutherAI/lm-evaluation-harness.

[8] https://huggingface.co/datasets/wikitext.

[9] https://medium.com/@NeuralCompressor/llm-performance-of-intel-extension-for-transformers-f7d061556176.

Table 3.2 Accuracy of INT4 and FP32 models. The INT4 model features two configurations with group sizes of 32 and 128

LLM	FP32	INT4 (Group size = 32)	INT4 (Group size = 128)
EleutherAI/gpt-j-6B	0.643	0.644	0.64
meta-llama/ Llama-2-7b-hf	0.69	0.69	0.685
decapoda-research/ llama-7b-hf	0.689	0.682	0.68
EleutherAI/ gpt-neox-20b	0.674	0.672	0.669
mosaicml/mpt-7b	0.689	0.688	0.683
tiiuae/falcon-7b	0.698	0.694	0.693
databricks/dolly-v2-3b	0.613	0.609	0.609
microsoft/phi-1.5	0.628	0.623	0.623
Qwen/Qwen-7B	0.683	0.68	0.672
mistralai/ Mistral-7B-v0.1	0.728	0.724	0.722

size of 128 allows for a greater reduction in memory usage than using a group size of 32.

Table 3.3 Memory usage (in MB) for INT4 and FP32 models. The INT4 model is available in two configurations with group sizes of 128 and 32

LLM	FP32	INT4 (Group size = 128)	INT4 (Group size = 32)
EleutherAI/gpt-j-6B	22481	3399	4017
meta-llama/ Llama-2-7b-hf	22057	3772	4928
decapoda-research/ llama-7b-hf	21750	3773	4874
EleutherAI/ gpt-neox-20b	68829	11221	13458
mosaicml/mpt-7b	21561	5678	5805
tiiuae/falcon-7b	24624	5335	5610
databricks/dolly-v2-3b	9389	2736	2794
microsoft/phi-1.5	5481	931	1059
Qwen/Qwen-7B	29715	4354	5245
mistralai/ Mistral-7B-v0.1	27835	4017	4975

3.4.4 Performance

We measure the latency of next token generation using LLM Runtime and the popular open-source GGML-based implementation. Table 3.4 presents the next token latency under a proxy configuration with 32 as both input and output token sizes. Note that the GGML-based solution only supports group size 32 when testing. As shown in the table, our solution achieves a 1.6x performance improvement compared to the GGML-based implementation. We marked the GGML-based dolly-v2-3b as 'N/A' because GGML-based implementation cannot perform inference with this model.

Table 3.5 presents the first token latency under a proxy configuration with 32 as both input and output token sizes. Due to MHA fusion, the latency for the first token is significantly lower than that of the GGML-based implementation.

We conducted another experiment on a client CPU, the Intel Core i9-12900, operating at 2.4GHz with 24 cores per socket and a total memory of 32GB (4× 8GB DDR5 4800 MT/s). The system was running BIOS version ADLSFWI1 R00.2257.A01.2106221245 with microcode 0x2e on Ubuntu 22.04.1 LTS. We benchmarked LLM Runtime using llama2-7b and llama.cpp with an input size of 1024, output size of 32, and beam size of 1. LLM Runtime showed a performance approximately three times faster compared to llama.cpp.

We also conducted experiments on tensor parallelism using an Intel Xeon 8481C processor running at 2.7GHz with 22 cores per socket. In these experiments, both the input size and output size were set to 32. The next token latency of the INT4

Table 3.4 INT4 next token latency using LLM Runtime and GGML-based solution (both in ms). LLM Runtime outperforms the GGML-based solution by up to 1.69x under group-size = 128, and by 1.52x under group size =3 2

Model	LLM Runtime (Group size = 32)	LLM Runtime (Group size = 128)	GGML-based (Group size = 32)
EleutherAI/gpt-j-6B	22.99	19.98	31.62
meta-llama/ Llama-2-7b-hf	23.4	21.96	27.71
decapoda-research/ llama-7b-hf	23.88	22.04	27.2
EleutherAI/ gpt-neox-20b	80.16	61.21	92.36
mosaicml/mpt-7b	23.04	21.05	31.54
tiiuae/falcon-7b	31.23	22.26	36.22
databricks/ dolly-v2-3b	15.3	11.66	N/A
microsoft/phi-1.5	10.44	7.25	15.88
Qwen/Qwen-7B	32.21	24.44	41.34
mistralai/ Mistral-7B-v0.1	29.87	23.85	39.9

Table 3.5 INT4 first token latency using LLM Runtime and GGML-based solutions (both in ms). LLM Runtime outperforms the GGML-based solution by up to 8.4x under group-size = 128, and by 3.75x under group size = 32

Model	LLM Runtime (Group size = 32)	LLM Runtime (Group size = 128)	GGML-based (Group size = 32)
EleutherAI/gpt-j-6B	64.32	43.93	390.6
meta-llama/ Llama-2-7b-hf	147.98	46.01	328.19
decapoda-research/ llama-7b-hf	103.33	57.65	324.85
EleutherAI/ gpt-neox-20b	362.52	161.97	1361.19
mosaicml/mpt-7b	259.67	62.95	321.31
tiiuae/falcon-7b	93.07	43.2	344.7
databricks/ dolly-v2-3b	63.83	38.64	N/A
microsoft/phi-1.5	64.99	19.78	96.11
Qwen/Qwen-7B	142.06	62.26	282.24
mistralai/ Mistral-7B-v0.1	154.42	115.3	317.44

Llama2-70b on 2 sockets is 200.6ms, and on a single socket, it is 387.6ms. This results in a scaling factor of 1.93x, indicating that the improvement nearly achieves linear growth with the number of sockets. The scaling factor for Llama2-13b is 1.83x, and for Llama2-7b, it is 1.8x. The scaling factor improves with larger model sizes. When we changed the input size to 1024, the scaling factor for INT4 Llama2-70b, Llama2-13b, and Llama2-7b increased to 1.89x, 1.84x, and 1.87x, respectively.

3.5 Conclusion and Future Work

We achieved up to 8.4x performance improvement in first-token processing and 1.69x in next-token processing, demonstrating a clear performance advantage over GGML-based solutions. At the same time, our solution still maintains good accuracy with advanced algorithms. Our solution presented an end-to-end INT4 LLM inference solution including automatic INT4 model quantization and an efficient LLM Runtime model. This solution supports a broad range of CPU platforms, including client CPUs, making it suitable for AI-powered machines, and meeting the growing demands for AI-generated content and enable generative AI on PCs. We demonstrated the generality of our approach on a set of popular LLMs and the performance advantage over the open-source solution on CPUs. Our solution has already extended HuggingFace Transformer APIs to support INT4 LLM inference as part of the contributions to the open-source community.[10]

There are still opportunities to enhance LLM Runtime performance through further tuning, such as optimizing the thread scheduler. While our threading currently relies on OpenMP, implementing a thread pool could potentially yield even better performance. Implementing a blocking strategy in the CPU tensor library would be another enhancement. Currently, we divide matrix multiplications using large block sizes, but using smaller sizes may improve performance for certain models.

As part of our future work, we plan to extend the solution to Intel GPUs and HPUs. Since we have tested tensor parallelism on multi-socket CPUs, GPUs can potentially achieve greater performance improvements. Meanwhile, we continue exploring quantization techniques and have published Autoround on GitHub.[11] We plan to explore 3-bit quantization, mixed precision, and 2-step quantization to provide the best quantization method that balances both accuracy and performance.

References

Bisk Y, Zellers R, Gao J, Choi Y et al (2020) Piqa: reasoning about physical commonsense in natural language. Proc AAAI Confer Artif Intell 34:7432–7439

Cheng W, Cai Y, Lv K, Shen H (2023a) Teq: trainable equivalent transformation for quantization of llms. arXiv preprint arXiv:2310.10944

Cheng W, Zhang W, Shen H, Cai Y, He X, Lv K (2023b) Optimize weight rounding via signed gradient descent for the quantization of llms. arXiv preprint arXiv:2309.05516

Dettmers T, Pagnoni A, Holtzman A, Zettlemoyer L (2023) Qlora: efficient finetuning of quantized llms. arXiv preprint arXiv:2305.14314

Frantar E, Ashkboos S, Hoefler T, Alistarh D (2022) Gptq: accurate post-training quantization for generative pre-trained transformers. arXiv preprint arXiv:2210.17323

Jiang AQ, Sablayrolles A, Mensch A, Bamford C, Chaplot DS, Casas Ddl, Bressand F, Lengyel G, Lample G, Saulnier L et al. (2023) Mistral 7b. arXiv preprint arXiv:2310.06825

Lin J, Tang J, Tang H, Yang S, Dang X, Han S (2023) Awq: activation-aware weight quantization for llm compression and acceleration. arXiv preprint arXiv:2306.00978

Micikevicius P, Stosic D, Burgess N, Cornea M, Dubey P, Grisenthwaite R, Ha S, Heinecke A, Judd P, Kamalu J et al. (2022) Fp8 formats for deep learning. arXiv preprint arXiv:2209.05433

Paperno D, Kruszewski G, Lazaridou A, Pham QN, Bernardi R, Pezzelle S, Baroni M, Boleda G, Fernández R (2016) The lambada dataset: word prediction requiring a broad discourse context. arXiv preprint arXiv:1606.06031

Rodriguez A, Segal E, Meiri E, Fomenko E, Kim YJ, Shen H, Ziv B (2018) Lower numerical precision deep learning inference and training. Intel White Paper 3(1):19

Rozière B, Gehring J, Gloeckle F, Sootla S, Gat I, Tan XE, Adi Y, Liu J, Remez T, Rapin J et al. (2023) Code llama: open foundation models for code. arXiv preprint arXiv:2308.12950

Sakaguchi K, Bras RL, Bhagavatula C, Choi Y (2021) Winogrande: an adversarial winograd schema challenge at scale. Commun ACM 64(9):99–106

[10] Users can easily select quantization approaches and inference parameters in https://github.com/intel/intel-extension-for-transformers.

[11] https://github.com/intel/auto-round.

Touvron H, Lavril T, Izacard G, Martinet X, Lachaux MA, Lacroix T, Rozière B, Goyal N, Hambro E, Azhar F et al. (2023a) Llama: open and efficient foundation language models. arXiv preprint arXiv:2302.13971

Touvron H, Martin L, Stone K, Albert P, Almahairi A, Babaei Y, Bashlykov N, Batra S, Bhargava P, Bhosale S et al. (2023b) Llama 2: open foundation and fine-tuned chat models. arXiv preprint arXiv:2307.09288

Vanhoucke V, Senior A, Mao MZ (2011) Improving the speed of neural networks on cpus. In: Deep learning and unsupervised feature learning workshop. NIPS 2011

Vaswani A, Shazeer NM, Parmar N, Uszkoreit J, Jones L, Gomez AN, Kaiser L, Polosukhin I (2017) Attention is all you need. In: Neural information processing systems. https://api.semanticscholar.org/CorpusID:13756489

Xiao G, Lin J, Seznec M, Wu H, Demouth J, Han S (2023) Smoothquant: accurate and efficient post-training quantization for large language models. In: International conference on machine learning, PMLR, pp 38087–38099

Zellers R, Holtzman A, Bisk Y, Farhadi A, Choi Y (2019) Hellaswag: can a machine really finish your sentence? arXiv preprint arXiv:1905.07830

Part III
Efficiency Techniques for Fine-Tuning

Chapter 4
KronA: Parameter-Efficient Tuning with Kronecker Adapter

Ali Edalati, Marzieh Tahaei, Ivan Kobyzev, Vahid Partovi Nia, James J. Clark, and Mehdi Rezagholizadeh

Abstract Fine-tuning a pre-trained language model (PLM) on a specific downstream task is a well-established paradigm in natural language processing. However, training the entire model on downstream tasks is computationally expensive, requiring significant time and resources. Parameter efficient fine tuning (PEFT) has been proposed to address this challenge by reducing the number of trainable parameters. A popular category of PEFT techniques inserts trainable adapters in a frozen-parameter model during the fine-tuning stage. Common adapters include low-rank projections such as LoRA, which reduces the adapter's representation power. We address this reduced representation using the Kronecker product instead of the low-rank projection to improve flexibility, leading to improved performance. We introduce KronA, a Kronecker equivalent of LoRA, to efficiently fine-tune Transformer-based PLMs. We apply different variants of KronA for fine-tuning the T5 model on the GLUE benchmark and show that KronA outperforms common PEFT baselines.

4.1 Introduction

Large pre-trained language models (PLMs) are used as a backbone in various natural language processing tasks to achieve state-of-the-art results (Devlin et al., 2019; Radford et al., 2019). PLMs are adapted to downstream applications either via in-context learning or fine-tuning. In-context learning imposes substantial memory and computational overhead during inference since all training examples have to be processed for each sample (Liu et al., 2022). In contrast, fine-tuning provides less inference latency and improved accuracy. However, as PLMs become larger, fine-tuning the entire model becomes challenging since more time and computational

A. Edalati (✉) · J. J. Clark
McGill University, Montreal, QC, Canada
e-mail: ali.edalati@mail.mcgill.ca

A. Edalati · M. Tahaei · I. Kobyzev · V. P. Nia · M. Rezagholizadeh
Huawei Noah's Ark Lab, Montreal, QC, Canada

© The Author(s), under exclusive license to Springer Nature Switzerland AG 2025
P. Passban et al. (eds.), *Enhancing LLM Performance*, Machine Translation: Technologies and Applications 7, https://doi.org/10.1007/978-3-031-85747-8_4

power is required. Additionally, one must store a complete model checkpoint for each downstream application, making the deployment inefficient.

To address these challenges, several works have proposed inserting just a few trainable parameters while freezing most (or even all) of the pre-trained model parameters. This significantly reduces the memory and computation requirements for fine-tuning. Furthermore, instead of storing one copy of the entire model, a small set of tuned parameters can be stored for each task. We refer to these methods as parameter-efficient fine tuning (PEFT) methods.

Among the PEFT methods, soft prompts (Lester et al., 2021; Li and Liang, 2021) prepend trainable parameters to the input of the layers and train them on downstream tasks. However, the increase in the length of the embedding layers leads to a significant computation overhead during inference.

In another category of the PEFT methods, trainable adapters (Houlsby et al., 2019; Mahabadi et al., 2021; He et al., 2022) are inserted into frozen Transformers (Vaswani et al., 2017). Adapters are low-rank modules that are composed of an up projection followed by a down projection. One limitation of these approaches is that they increase the computational overhead and the latency during inference which makes them inefficient for latency-critical scenarios.

Accordingly, low-rank adaption (LoRA) (Hu et al., 2022) was developed using extra low-rank adapters in parallel to pre-trained weight matrices. However, once fine-tuned, the adapter parameters can be merged with the original pre-trained weights, making the latency and energy requirements for inference, intact. Despite the fast inference, like other low-rank adapters, LoRA suffers from a loss of accuracy compared to full fine-tuning. This is due to the strong assumption imposed by its low-rank structure for the task-specific updates.

Kronecker-based decomposition is another factorization method that does not rely on the low-rank assumption. When used for model compression, this powerful decomposition method has proven to outperform low-rank compression methods (Thakker et al., 2019; Hameed et al., 2022). It has also been used successfully to compress Transformer-based language models (Edalati et al., 2022; Tahaei et al., 2022).

Inspired by the success of Kronecker decomposition, we replace the low-rank projections of LoRA with the Kronecker product to develop Kronecker adapter (KronA). This simple modification can improve accuracy without increasing the inference latency. Furthermore, for applications where the latency increase is tolerable, we propose to use KronA^{B}. This module is a version of KronA developed to be utilized in parallel to feedforward network (FFN) blocks and achieves notable improvements over full fine-tuning on the general language understanding evaluation (GLUE) benchmark (Wang et al., 2019). In addition, when a proposed learnable residual connection is added to KronA^{B}, $\text{KronA}^{\text{B}}_{\text{res}}$ is developed to achieve even better results.

We evaluated our methods on the GLUE benchmark to study the impact of the Kronecker product on the performance. To summarize, our contributions are:

- Proposing KronA, a Kronecker Adapter that can be inserted in parallel to the weight matrices and is suitable for latency-critical scenarios.
- Using KronA in parallel to FFNs (KronA^{B}) along with a learnable residual connection (KronA$^{B}_{res}$) to further improve accuracy at the cost of a higher inference latency.
- Providing evaluation of the proposed methods in comparison to the well-known baselines in terms of the GLUE score, training time, and inference latency.

4.2 Related Work

BitFit (Ben Zaken et al., 2022) proposed freezing the weights and tuning only (a subset of) biases of PLMs for fine-tuning on downstream tasks. This technique is parameter-efficient and fast, but it underperforms compared to state-of-the-art methods.

Adapters (Houlsby et al., 2019) were introduced as a PEFT method, where all parameters of a PLM are frozen and some trainable modules are inserted after the FFN or attention blocks. The proposed adapters contain a down projection, a non-linear function, an up projection, and a residual connection. We use "Adapter" to refer to these modules. Later, parallel adapters (PAs) (He et al., 2022) were developed as another version of adapters. PAs are inserted parallel to the pre-trained blocks and have a scaling factor (Fig. 4.1c). In addition, a unified view on PEFT methods and combinations of some techniques such as prefix tuning (Li and Liang, 2021) and PA was studied in He et al. (2022).

Compacter (Mahabadi et al., 2021) proposed a modified version of the adapters for PEFT. In Compacter, the Kronecker product of multiple pairs of Kronecker factors is summed to reconstruct the module's weight matrix ($\mathbf{W}_{\text{Compacter}} = \sum_{i=1}^{n} \mathbf{A}_i \otimes \mathbf{B}_i$). To further reduce the trainable parameters, $\mathbf{A}_i$ matrices are shared across all layers and $\mathbf{B}_i$ matrices are decomposed as the matrix multiplication of two low-rank sub-factors. Although Compacter achieves good results, it is notably slow in the training and inference phases.

KAdaptation (He et al., 2023) developed Kronecker-based adapters that have a similar structure to Compacter adapters for Vision Transformers (Dosovitskiy et al., 2021). However, KAdaptation adapters are applied in parallel to weight matrices which enables merging the adapters into the model after training to avoid increasing the latency and parameters at the inference stage. Furthermore, PEFT of both convolutional and linear layers used in computer vision models was investigated in Edalati et al. (2023) by developing adapters that use the summation of the Kronecker product of a sequence of factors.

The proposed adapters in our work have a simpler structure compared to those by Mahabadi et al. (2021), Edalati et al. (2023), and He et al. (2023). By eliminating parameter sharing, low-rank sub-factor decomposition, and summation, we have developed a faster method, albeit with less parameter reduction

LoRA (Hu et al., 2022) inserted adapters made from a downward projection and an upward projection parallel to the PLM weight matrices to introduce LoRA. During the training, the pre-trained weights are frozen, and only the LoRA adapters are tuned. During inference, the LoRA adapters are merged with the original weight matrices of PLMs. Therefore, unlike Adapter, PA, and Compacter, LoRA does not increase inference time.

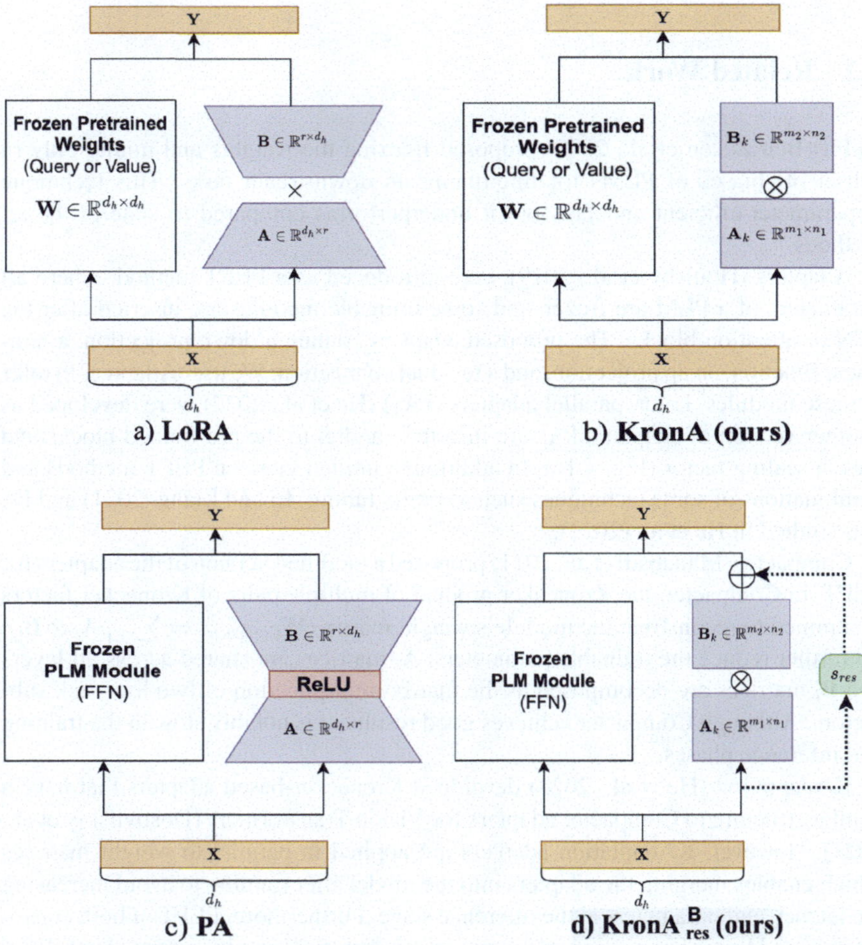

Fig. 4.1 The structure of the proposed Kronecker-based adapters and their low-rank counterparts. For simplicity, the scaling factor at the output of the modules is not depicted. Figure **d** shows $\text{KronA}^{\text{B}}_{\text{res}}$. The residual connection is depicted by a dotted-line to indicate that this connection can simply be removed to have KronA^{B}

4.3 Methodology

In this section, the Kronecker product, its features, and our approach for developing the Kronecker-based adapters are explained. For more details see Edalati (2023).

4.3.1 Kronecker Product

Equation (4.1) shows how the elements of **A** are multiplied by **B** to generate the Kronecker product of $\mathbf{A} \in \mathbb{R}^{m_1 \times n_1}$ and $\mathbf{B} \in \mathbb{R}^{m_2 \times n_2}$ (Henderson et al., 1983).

$$\mathbf{A} \otimes \mathbf{B} = \begin{bmatrix} a_{1,1}\mathbf{B} & \cdots & a_{1,n_1}\mathbf{B} \\ \vdots & \ddots & \vdots \\ a_{m_1,1}\mathbf{B} & \cdots & a_{m_1,n_1}\mathbf{B} \end{bmatrix} = \mathbf{W} \in \mathbb{R}^{m_1 m_2 \times n_1 n_2} \tag{4.1}$$

The Kronecker product has some interesting features that make it suitable for PEFT. First, unlike the low-rank down-projections used in other techniques, Kronecker-based decomposition maintains the rank of the input matrix. Second, to reduce the required FLOPS, Eq. (4.2) can be used to obtain $(\mathbf{A} \otimes \mathbf{B})\mathbf{x}$ without the computation of $\mathbf{A} \otimes \mathbf{B}$, where $\eta_{m \times n}(\cdot)$ reshapes a vector into a matrix of size $m \times n$ and $\gamma(\cdot)$ flattens a matrix into a vector.

$$(\mathbf{A} \otimes \mathbf{B})\mathbf{x} = \gamma(\mathbf{B}\eta_{m \times n}(\mathbf{x})\mathbf{A}^\top) \tag{4.2}$$

4.3.2 KronA

Figure 4.1a shows the structure of a LoRA adapter where **A** and **B** are the weight matrices of the down and up projection, respectively. To modify this module into KronA, the Kronecker product replaces the matrix multiplication. In addition, the LoRA projections are replaced by Kronecker factors (see Fig. 4.1b). Table 4.1 shows the features of the Kronecker factors that replaced the projections in the LoRA modules. Equation (4.3) shows how the output is generated when KronA is applied. $\mathbf{A}_k$ and $\mathbf{B}_k$ are the Kronecker factors that replaced the LoRA projections. Similar to LoRA, KronA has a fixed scaling factor, s, which is a hyperparameter.

$$\mathbf{Y} = \mathbf{X}\mathbf{W} + s\mathbf{X}[\mathbf{A}_k \otimes \mathbf{B}_k] \tag{4.3}$$

The KronA modules are applied in parallel to the weight matrices of PLMs during the fine-tuning phase. Once fine-tuned, the Kronecker factors are multiplied, then

Table 4.1 Comparing some details of the Kronecker factors with the LoRA projections

Module name	Factor name	Symbol	Shape	Parameters	Module parameters	Constraint
KronA	Kronecker factor	$\mathbf{A}_k$	$m_1 \times n_1$	$m_1 n_1$	$m_1 n_1 + m_2 n_2$	$m_1 m_2 = n_1 n_2 = d_h$
	Kronecker factor	$\mathbf{B}_k$	$m_2 \times n_2$	$m_2 n_2$		
LoRA	Down projection	$\mathbf{A}$	$d_h \times r$	$d_h r$	$2 d_h r$	$r < \frac{d_h}{2}$
	Up projection	$\mathbf{B}$	$r \times d_h$	$d_h r$		

scaled and merged into the original weight matrices (Eq. 4.4). Therefore, similar to LoRA, KronA does not increase the inference time.

$$\mathbf{W}_{\text{tuned}} = \mathbf{W} + s[\mathbf{A}_k \otimes \mathbf{B}_k] \tag{4.4}$$

4.3.3 $KronA^B$ and $KronA^B_{res}$

Inspired by the promising performance of PA, we also investigate KronA when used in parallel to the FNN blocks and call it KronA^{B}. The B superscript in the name means that this module is applied to the pre-trained blocks, as opposed to KronA, which is applied to pre-trained weight matrices. Similar to PA, the non-linearity in the FFN blocks does not allow our proposed adapters to be merged into the pre-trained blocks after fine-tuning. This imposes an increase in the inference time and computations. Equation (4.5) shows how KronA^{B} works in parallel to an FFN block.

$$\mathbf{Y} = \text{FFN}(\mathbf{X}) + s\mathbf{X}[\mathbf{A}_k \otimes \mathbf{B}_k] \tag{4.5}$$

To further improve the representation power, we incorporate a scaled residual connection inside the KronA^{B} module to develop $\text{KronA}^{\text{B}}_{\text{res}}$. The scale of the residual connection (s_{res}) is initialized with one and tuned during the fine-tuning. Equation (4.6) shows how $\text{KronA}^{\text{B}}_{\text{res}}$ works in parallel to an FFN block. Furthermore, Fig. 4.1c and d show the structure of a PA and our $\text{KronA}^{\text{B}}_{\text{res}}$ module, respectively.

$$\mathbf{Y} = \text{FFN}(\mathbf{X}) + s\mathbf{X}[\mathbf{A}_k \otimes \mathbf{B}_k] + s_{\text{res}}\mathbf{X} \tag{4.6}$$

Table 4.2 The performance of the tested options for $\mathbf{A}_k$ in KronA on MNLI. Note that for each option, the shape of the corresponding $\mathbf{B}_k$ is in the reversed order of $\mathbf{A}_k$

Shape of $\mathbf{A}_k$	MNLI (accuracy)
(48, 16)	86.50
(32, 24)	86.31
(3, 256)	86.16
(24, 32)	86.40
(2, 384)	**86.63**
(192, 4)	86.46
(12, 64)	86.56

4.4 Experimental Setup

This section provides information about our experimental setup including the hyperparameters, datasets, implementation, and time measurement.

4.4.1 Setting and Hyperparameters

All experiments were performed on one NVIDIA Tesla V100. We used PyTorch and the HuggingFace Transformers library (Wolf et al., 2020) for our experiments. To re-implement LoRA[1] and PA,[2] we used their publicly available code. For the experiments on Compacter, BitFit, fine-tuning, and Adapter, we used the official Compacter code.[3] The backbone model for this work is T5_{base} (Raffel et al., 2020). The size of the trainable parameters for all of the methods is set roughly equal to enable a fair comparison. However, for BitFit tuning, we could not match the trainable parameters despite tuning all of the biases.

Given the number of trainable parameters, we have several choices for the shape of the Kronecker factors. For KronA, we tested some of the options and selected the one option with the best results (see Table 4.2). KronA^{B} and $\text{KronA}^{\text{B}}_{\text{res}}$ modules can have one or two biases. We selected the number of biases that maximized the score on each task. Since we wanted to ignore the effect of the scaling factor when comparing LoRA and KronA, the scaling factor for these two modules is set to one in all experiments.

For fine-tuning, BitFit, Compacter, and Adapter experiments, we used the hyperparameters that are mentioned in Mahabadi et al. (2021). However, we changed the learning rate and the rank of the modules to match the desired number of trainable parameters in the Adapter experiments.

---

[1] https://github.com/microsoft/LoRA.

[2] https://github.com/jxhe/unify-parameter-efficient-tuning.

[3] https://github.com/rabeehk/compacter.

Table 4.3 The hyperparameters used for the experiments

Fine-tuning and BitFit hyperparameters

Task	Learning rate	Batch size	Warmup steps	Source sentence length	Epoch
GLUE	3e-4	100	500	128	20

Adapter hyperparameters

Task	Learning rate	Task reduction factor
GLUE	3e-3	32

LoRA hyperparameters

Learning rate	Rank	s
1e-3	1	1

Compacter hyperparameters

Task	Learning rate	Hypercomplex division	Task reduction factor
GLUE	3e-3	4	32

	PA hyperparameters			$KronA^B_{res}$ hyperparameters				
Task	Learning rate	rank	s	Learning rate	$\mathbf{A}_k$	$\mathbf{B}_k$	s	Module bias
CoLA	3e-3	2	16	1e-3	(32, 24)	(24, 32)	16	2
RTE	5e-3	2	16	5e-3	(32, 24)	(24, 32)	16	2
MRPC	5e-3	2	16	5e-3	(32, 24)	(24, 32)	16	2
SST-2	1e-3	2	16	1e-3	(32, 24)	(24, 32)	16	1
STS-B	1e-3	2	16	9e-4	(32, 24)	(32, 24)	16	1
MNLI	1e-3	2	16	1e-3	(32, 24)	(24, 32)	16	1
QNLI	1e-3	2	16	1e-3	(32, 24)	(24, 32)	4	2
QQP	1e-3	2	16	1e-3	(32, 24)	(24, 32)	16	1

(continued)

Table 4.3 (continued)

Fine-tuning and BitFit hyperparameters

KronA hyperparameters					*KronA*^{B} hyperparameters				
Task	Learning rate	$\mathbf{A}_k$	$\mathbf{B}_k$	*s*	Learning rate	$\mathbf{A}_k$	$\mathbf{B}_k$	*s*	Module bias
CoLA	1e-3	(32, 24)	(24, 32)	1	1e-3	(32, 24)	(24, 32)	16	2
RTE	2e-3	(32, 24)	(24, 32)	1	5e-3	(32, 24)	(24, 32)	16	1
MRPC	1e-3	(32, 24)	(24, 32)	1	5e-3	(32, 24)	(24, 32)	16	1
SST-2	1e-3	(24, 32)	(32, 24)	1	1e-3	(32, 24)	(24, 32)	4	1
STS-B	1e-3	(2, 384)	(384, 2)	1	1e-3	(32, 24)	(32, 24)	16	1
MNLI	1e-3	(2, 384)	(384, 2)	1	1e-3	(32, 24)	(24, 32)	4	2
QNLI	1e-3	(3, 256)	(256, 3)	1	1e-3	(32, 24)	(24, 32)	4	1
QQP	1e-3	(24, 32)	(32, 24)	1	1e-3	(32, 24)	(24, 32)	4	1

The rank of LoRA and PA is set to one and two, respectively. For the KronA modules, the shape of the Kronecker factors is selected based on the best dev results among different options for the shapes. Due to time and resource limitations, we did not tune the shape of the Kronecker factors for KronA^{B} and $\text{KronA}^{\text{B}}_{\text{res}}$.

All of the other hyperparameters are set based on Mahabadi et al. (2021), except for the learning rate and scaling factor which are tuned based on the best dev results. All methods are trained for 20 epochs and the checkpoint that achieves the best performance on the dev set is reported as the final model. Table 4.3 shows the tuned hyperparameters for each method on the GLUE tasks.

4.4.2 Datasets

We used the GLUE benchmark to evaluate our methods compared to the baselines. This benchmark covers a variety of tasks including:

- Natural language inference: MNLI (Williams et al., 2018), RTE (Bar-Haim et al., 2006; Bentivogli et al., 2009; Dagan et al., 2006; Giampiccolo et al., 2007), QNLI (Rajpurkar et al., 2016)
- linguistic acceptability: CoLA (Warstadt et al., 2019)
- similarity and paraphrasing: STS-B (Cer et al., 2017), MRPC (Dolan and Brockett, 2005), QQP[4]
- sentiment classification: SST-2 (Socher et al., 2013)

The original GLUE test labels are not published, so similar to Mahabadi et al. (2021); Zhang et al. (2023), we generated our test sets from the evaluation and the training data. For the small datasets (CoLA, RTE, MRPC, and STS-B), we used half of the task dev set for evaluation and the other half as the test set. For the rest of the GLUE tasks with larger datasets, we took 1K samples from the training set and used them as our test sets. The reported evaluation metric for CoLA, MRPC, and STS-B is the Matthew correlation coefficient (Matthews, 1975), F1, and the average of Pearson/Spearman correlations, respectively. Accuracy is used for the other tasks.

4.4.3 Details of Measuring the Training and Inference Time

To measure the inference latency, a random dummy input with a batch size equal to one and a sequence length equal to ten is generated. Then, the dummy input is given to the model for 150 iterations to warm up the GPU. Finally, the dummy input is fed to the model for 200 iterations and the required time to generate the output is measured, averaged, and recorded. This experiment is repeated three times and the

[4] https://data.quora.com/First-Quora-Dataset-Release-Question-Pairs.

Table 4.4 The normalized training time of methods on the GLUE tasks

Method	CoLA	RTE	MRPC	SST-2	STS-B	MNLI	QNLI	QQP	Average
BitFit	0.58	0.64	0.65	0.66	0.66	0.65	0.64	0.62	0.64
Adapter	0.82	0.71	0.72	0.78	0.76	0.72	0.69	0.69	0.73
LoRA	0.79	0.69	0.7	0.76	0.72	0.7	0.68	0.68	0.72
KronA	0.8	0.72	0.75	0.81	0.77	0.74	0.73	0.73	0.75
Compacter	0.88	0.74	0.78	0.86	0.81	0.75	0.74	0.76	0.79
PA	0.7	0.78	0.81	0.73	0.7	0.67	0.66	0.65	0.71
KronA^{B}	0.84	0.7	0.72	0.79	0.75	0.72	0.7	0.71	0.74
$\text{KronA}^{\text{B}}_{\text{res}}$	0.91	0.85	0.81	0.91	0.76	0.78	0.73	0.74	0.81

average latency is reported. Finally, the reported latencies are normalized and shown in Table 4.6.

Table 4.4 shows the normalized training time for each technique on the GLUE tasks. All the experiments are performed with the same number of epochs, batch size, number of GPUs, and the gradient accumulation step.

4.5 Results and Discussion

Table 4.5 shows the GLUE score of our proposed adapters compared to other baselines when applied to T5 (Raffel et al., 2020). As the results show, KronA and KronA^{B} outperform LoRA and PA as their low-rank counterparts, respectively. All of our proposed adapters also outperform other baselines on average and most of the GLUE tasks. Furthermore, $\text{KronA}^{\text{B}}_{\text{res}}$, which benefits from an extra learnable residual connection, achieves remarkably better results.

Table 4.6 shows the normalized inference delay for the discussed methods. KronA, LoRA, fine-tuning, and Bitfit do not increase the inference latency since these techniques do not add additional parameters or computations to the model during the inference phase. Although KronA^{B} is significantly faster than Compacter and Adapter, it is slower than PA, as expected; computation of the Kronecker product is generally slower than the normal matrix multiplication. Also, adding the learnable residual connection increases the latency.

The normalized training time averaged over the GLUE tasks for each technique is shown in Table 4.6. Based on these results, the significant improvement in the accuracy of the proposed KronA and its variants is at the expense of a slight increase in the training time compared to the low-rank counterparts like LoRA, PA, and Adapter. However, the training time increase is not remarkable, and KronA modules are still significantly faster than fine-tuning.

Table 4.5 The score of the methods on GLUE

Method	Parameters	CoLA	RTE	MRPC	SST-2	STS-B	MNLI	QNLI	QQP	Average
Fine-tuning	100	63.37	74.82	92.73	93.58	90.07	86.16	92.77	91.74	85.65
BitFit	0.12	58.19	68.34	92.58	94.61	90.69	85.73	92.91	90.33	84.17
Adapter	0.07	64.66	71.94	91.27	**94.84**	90.49	85.91	92.97	90.35	85.30
LoRA	0.07	64.76	74.10	92.10	93.92	91.21	86.08	92.97	**90.68**	85.73
Compacter	0.07	64.42	76.26	91.52	93.92	91.04	86.14	92.93	90.36	85.82
PA	0.06	64.80	74.10	**93.20**	94.04	91.10	86.24	93.12	90.30	85.86
KronA	0.07	63.27	**77.70**	92.52	94.26	91.30	**86.34**	93.15	90.57	**86.14**
KronA^{B}	0.07*	65.74	75.54	92.78	94.72	**91.41**	86.22	93.19	**90.68**	**86.28**
$\text{KronA}^{\text{B}}_{\text{res}}$	0.07*	**66.73**	76.98	93.15	94.38	91.35	86.20	**93.21**	90.57	**86.57**

$*$ shows that the number of trainable parameters might be slightly different depending on the choice of one or two biases in the module

Table 4.6 The first row shows the normalized latency of the methods in the inference phase while the second row lists the average normalized training time of the methods on the GLUE tasks

Method	Fine-tuning	LoRA	KronA	BitFit	Adapter	PA	Compacter	KronA^{B}	$\text{KronA}^{\text{B}}_{\text{res}}$
Inference Latency (%)	100	100	100	100	146	113	181	127	136
Training Time (%)	100	72	75	64	73	71	79	74	81

4.6 Ablation Study

This section provides an ablation study on our work, illustrating the effect of the steps that were taken to develop our proposed Kronecker-based adapters.

4.6.1 KronA Initialization

Our empirical results show that the initialization of the Kronecker factors affects the performance of KronA. Table 4.7 shows the performance of two investigated strategies for the initialization. We observe that by initializing one of the Kronecker factors from a Kaiming-uniform (KU) distribution ($a = \sqrt{5}$) (He et al., 2015) and the other one with zero, KronA adapters perform significantly better than initializing both of the factors from a Normal ($\mu = 0, \sigma = \frac{1}{\sqrt{d_h}}$, where d_h is the embedding dimension) distribution.

Table 4.7 The performance of KronA on GLUE using different initialization options for the Kronecker factors. In the first row, both factors are initialized from a normal distribution ($\mu = 0, \sigma = \frac{1}{\sqrt{d_h}}$) while in the second row, $\mathbf{A}_k$ is initialized from a Kaiming-Uniform distribution ($a = \sqrt{5}$) and $\mathbf{B}_k$ is set to zero

Init Method	CoLA	RTE	MRPC	SST-2	STS-B	MNLI	QNLI	QQP	Average
$\mathbf{A}_k$,$\mathbf{B}_k$ ∼ Normal	63.36	66.91	91.69	91.97	90.46	86.03	92.33	90.19	84.12
$\mathbf{A}_k$ ∼ KU ,$\mathbf{B}_k$=0	63.27	77.70	92.52	94.04	91.26	86.03	93.13	90.57	86.06

4.6.2 Step by Step Improvement of KronA^B

At first, KronA^B was initialized like a normal adapter. It was sequentially inserted after both FFN and attention blocks and it did not have a scaling factor. We modified our module based on He et al. (2022) to improve its performance.

Table 4.8 shows the results of our experiments. We observed that inserting KronA^B modules in parallel to the PLM modules, rather than sequentially inserting them, significantly improves the performance. Additionally, adding a scaling factor to our module further increases the GLUE score. Furthermore, adding two modules to each FFN instead of adding to both the FFN and the attention blocks resulted in a higher score.

In addition, motivated by the presence of a non-linear function in PA and Adapter, we tested different non-linear functions between the two multiplications (by $\mathbf{A}_k^T$ and $\mathbf{B}_k$) in Eq. (4.2). As Table 4.9 shows, SiLU (Elfwing et al., 2018a) is the best option among others, but according to Table 4.8, adding SiLU decreases the GLUE score of KronA^B. Therefore, we removed the nonlinearity from our module.

Table 4.8 The performance of KronA^B after implementing step by step modifications on GLUE

Modification	CoLA	RTE	MRPC	SST-2	STS-B	MNLI	QNLI	QQP	Avg
Sequential	15.26	53.28	86.39	87.38	83.78	74.70	84.29	86.63	71.46
Parallel	58.17	69.78	91.58	93.81	90.86	85.68	93.35	90.14	84.17
Parallel+Scale (PS)	62.27	70.50	91.58	94.04	91.01	86.16	93.39	90.61	84.94
PS+SiLU	62.74	69.78	91.89	94.15	90.97	85.98	93.30	90.15	84.87
PS only on FFN	63.74	72.66	92.20	94.72	90.98	85.98	93.12	90.68	85.51

Table 4.9 The performance of KronA^B on QNLI using Mish (Misra, 2020), ReLU (Agarap, 2018), GELU (Hendrycks and Gimpel, 2016), GELU$_{new}$, and SiLU (Elfwing et al., 2018b) as different non-linear functions

Nonlinear function	Mish	ReLU	GELU	GELU$_{new}$	SiLU
QNLI Performance	93.21	93.28	93.13	93.26	93.30

Table 4.10 The effect of adding a sigmoid function to the $\text{KronA}^{\text{B}}_{\text{res}}$ module. "Avg Score" is the averaged score on the GLUE tasks and "Training Time" represents the relative training time

Method	Average score	Training time
$\text{KronA}^{\text{B}}_{\text{res}}$	**86.57**	1
$\text{KronA}^{\text{B}}_{\text{sigres}}$	86.42	1.18

4.6.3 Learnable Residual Connection

In the $\text{KronA}^{\text{B}}_{\text{res}}$ module, a residual connection multiplied by a learnable scale is added to the output of KronA^{B}. We also studied another scenario in which the learnable scale is passed through a sigmoid function and then multiplied by the residual connection. This module is called $\text{KronA}^{\text{B}}_{\text{sigres}}$. We wanted to investigate whether it was better to limit the residual scale between 0 and 1. Our empirical results (Table 4.10) show that by adding the sigmoid function, the performance of the module drops and the latency increases. Therefore, the sigmoid function was removed from our module.

4.7 Conclusion

In this chapter, we proposed Kronecker-based adapters by replacing the low-rank projections from well-known PEFT methods with the Kronecker product. In addition to comparing the training and inference time, we evaluated our proposed adapter for fine-tuning T5 on the GLUE benchmark to show its superiority over common baselines.

References

Agarap AF (2018) Deep learning using rectified linear units (relu). arXiv preprint arXiv:1803.08375

Bar-Haim R, Dagan I, Dolan B, Ferro L, Giampiccolo D, Magnini B, Szpektor I (2006) The second pascal recognising textual entailment challenge. In: Proceedings of the second PASCAL challenges workshop on recognising textual entailment, vol 1

Ben Zaken E, Goldberg Y, Ravfogel S (2022) BitFit: Simple parameter-efficient fine-tuning for transformer-based masked language-models. In: Proceedings of the 60th annual meeting of the association for computational linguistics (Volume 2: Short papers), Dublin, Ireland, pp 1–9. https://aclanthology.org/2022.acl-short.1

Bentivogli L, Magnini B, Dagan I, Dang HT, Giampiccolo D (2009) The fifth PASCAL recognizing textual entailment challenge. In: Proceedings of the second text analysis conference, TAC 2009, Gaithersburg, Maryland, USA, November 16–17, 2009, NIST

Cer D, Diab M, Agirre E, Lopez-Gazpio I, Specia L (2017) SemEval-2017 task 1: semantic textual similarity multilingual and crosslingual focused evaluation. In: Proceedings of the 11th international workshop on semantic evaluation (SemEval-2017), Vancouver, Canada, pp 1–14

Dagan I, Glickman O, Magnini B (2006) The PASCAL recognising textual entailment challenge. In: Machine learning challenges. evaluating predictive uncertainty, visual object classification, and recognising textual entailment. Springer, Berlin, pp 177–190

Devlin J, Chang MW, Lee K, Toutanova K (2019) BERT: pre-training of deep bidirectional transformers for language understanding. In: Proceedings of the 2019 conference of the North American chapter of the association for computational linguistics (ACL): human language technologies, vol 1 (Long and Short Papers), pp 4171–4186

Dolan WB, Brockett C (2005) Automatically constructing a corpus of sentential paraphrases. In: Proceedings of the third international workshop on paraphrasing (IWP2005)

Dosovitskiy A, Beyer L, Kolesnikov A, Weissenborn D, Zhai X, Unterthiner T, Dehghani M, Minderer M, Heigold G, Gelly S, Uszkoreit J, Houlsby N (2021) An image is worth 16x16 words: transformers for image recognition at scale. In: International conference on learning representations (ICLR)

Edalati A (2023) The Kronecker product for efficient natural language processing. Master's thesis, McGill University (Canada)

Edalati A, Tahaei M, Rashid A, Nia V, Clark J, Rezagholizadeh M (2022) Kronecker decomposition for GPT compression. In: Proceedings of the 60th annual meeting of the association for computational linguistics (Volume 2: Short papers), Dublin, Ireland, pp 219–226. https://aclanthology.org/2022.acl-short.24

Edalati A, Abdel Hameed MG, Mosleh A (2023) Generalized Kronecker-based adapters for parameter-efficient fine-tuning of vision transformers. In: 2023 20th conference on robots and vision (CRV), pp 97–104

Elfwing S, Uchibe E, Doya K (2018) Sigmoid-weighted linear units for neural network function approximation in reinforcement learning. Neural Netw: Official J Int Neural Netw Soc 107:3–11

Elfwing S, Uchibe E, Doya K (2018) Sigmoid-weighted linear units for neural network function approximation in reinforcement learning. Neural Netw: Official J Int Neural Netw Soc 107:3–11

Giampiccolo D, Magnini B, Dagan I, Dolan B (2007) The third PASCAL recognizing textual entailment challenge. In: Proceedings of the ACL-PASCAL workshop on textual entailment and paraphrasing, Prague, pp 1–9

Hameed MGA, Tahaei MS, Mosleh A, Nia VP (2022) Convolutional neural network compression through generalized Kronecker product decomposition. In: Thirty-Sixth AAAI conference on artificial intelligence, AAAI 2022, thirty-fourth conference on innovative applications of artificial intelligence, IAAI 2022, the twelveth symposium on educational advances in artificial intelligence, EAAI 2022 Virtual Event, February 22 - March 1, 2022. AAAI Press, pp 771–779

He J, Zhou C, Ma X, Berg-Kirkpatrick T, Neubig G (2022) Towards a unified view of parameter-efficient transfer learning. In: International conference on learning representations (ICLR)

He K, Zhang X, Ren S, Sun J (2015) Delving deep into rectifiers: Surpassing human-level performance on imagenet classification. In: Proceedings of the IEEE international conference on computer vision, pp 1026–1034

He X, Li C, Zhang P, Yang J, Wang XE (2023) Parameter-efficient model adaptation for vision transformers. Proc AAAI Conf Artif Intell 37(1):817–825

Henderson HV, Pukelsheim F, Searle SR (1983) On the history of the Kronecker product. Linear Multilinear Algebra 14:113–120

Hendrycks D, Gimpel K (2016) Gaussian error linear units (gelus). arXiv preprint arXiv:1606.08415

Houlsby N, Giurgiu A, Jastrzebski S, Morrone B, De Laroussilhe Q, Gesmundo A, Attariyan M, Gelly S (2019) Parameter-efficient transfer learning for nlp. In: International conference on machine learning, PMLR, pp 2790–2799

Hu EJ, yelong shen, Wallis P, Allen-Zhu Z, Li Y, Wang S, Wang L, Chen W (2022) LoRA: low-rank adaptation of large language models. In: International conference on learning representations (ICLR). https://openreview.net/forum?id=nZeVKeeFYf9

Lester B, Al-Rfou R, Constant N (2021) The power of scale for parameter-efficient prompt tuning. In: Proceedings of the 2021 conference on empirical methods in natural language processing,

Online and Punta Cana, Dominican Republic, pp 3045–3059. https://aclanthology.org/2021.emnlp-main.243

Li XL, Liang P (2021) Prefix-tuning: Optimizing continuous prompts for generation. In: Proceedings of the 59th annual meeting of the association for computational linguistics and the 11th international joint conference on natural language processing (vol 1: Long Papers), pp 4582–4597. https://aclanthology.org/2021.acl-long.353

Liu H, Tam D, Muqeeth M, Mohta J, Huang T, Bansal M, Raffel CA (2022) Few-shot parameter-efficient fine-tuning is better and cheaper than in-context learning. In: Advances in neural information processing systems. Curran Associates, Inc., vol 35, pp 1950–1965. https://proceedings.neurips.cc/paper_files/paper/2022/file/0cde695b83bd186c1fd456302888454c-Paper-Conference.pdf

Mahabadi RK, Henderson J, Ruder S (2021) Compacter: efficient low-rank hypercomplex adapter layers. In: Advances in neural information processing systems 34: annual conference on neural information processing systems 2021, NeurIPS, pp 1022–1035. https://dblp.org/rec/conf/nips/MahabadiHR21.bib

Matthews B (1975) Comparison of the predicted and observed secondary structure of t4 phage lysozyme. Biochimica et Biophysica Acta (BBA) - Protein Struct 405(2):442–451

Misra D (2020) Mish: a self regularized non-monotonic activation function. In: 31st British machine vision conference 2020, BMVC 2020, Virtual Event, UK, September 7–10, 2020. BMVA Press

Radford A, Wu J, Child R, Luan D, Amodei D, Sutskever I et al (2019) Language models are unsupervised multitask learners, p 9

Raffel C, Shazeer N, Roberts A, Lee K, Narang S, Matena M, Zhou Y, Li W, Liu PJ (2020) Exploring the limits of transfer learning with a unified text-to-text transformer. J Mach Learn Res 21(1):5485–5551

Rajpurkar P, Zhang J, Lopyrev K, Liang P (2016) SQuAD: 100,000+ questions for machine comprehension of text. In: Proceedings of the 2016 conference on empirical methods in natural language processing, Austin, Texas, pp 2383–2392

Socher R, Perelygin A, Wu J, Chuang J, Manning CD, Ng A, Potts C (2013) Recursive deep models for semantic compositionality over a sentiment treebank. In: Proceedings of the 2013 conference on empirical methods in natural language processing, Seattle, Washington, USA, pp 1631–1642

Tahaei M, Charlaix E, Nia V, Ghodsi A, Rezagholizadeh M (2022) KroneckerBERT: significant compression of pre-trained language models through Kronecker decomposition and knowledge distillation. In: Proceedings of the 2022 conference of the North American chapter of the association for computational linguistics: human language technologies, Seattle, United States, pp 2116–2127. https://aclanthology.org/2022.naacl-main.154

Thakker U, Fedorov I, Beu J, Gope D, Zhou C, Dasika G, Mattina M (2019) Pushing the limits of rnn compression. In: 2019 fifth workshop on energy efficient machine learning and cognitive computing-NeurIPS edition (EMC2-NIPS). IEEE, pp 18–21

Vaswani A, Shazeer NM, Parmar N, Uszkoreit J, Jones L, Gomez AN, Kaiser L, Polosukhin I (2017) Attention is all you need. In: Neural information processing systems. https://api.semanticscholar.org/CorpusID:13756489

Wang A, Singh A, Michael J, Hill F, Levy O, Bowman SR (2019) GLUE: a multi-task benchmark and analysis platform for natural language understanding. In: 7th international conference on learning representations, ICLR 2019, New Orleans, LA, USA, May 6–9, 2019, OpenReview.net

Warstadt A, Singh A, Bowman SR (2019) Neural network acceptability judgments. Trans Assoc Comput Linguist 7:625–641

Williams A, Nangia N, Bowman S (2018) A broad-coverage challenge corpus for sentence understanding through inference. In: Proceedings of the 2018 conference of the North American chapter of the association for computational linguistics: human language technologies, vol 1 (Long papers), New Orleans, Louisiana, pp 1112–1122

Wolf T, Debut L, Sanh V, Chaumond J, Delangue C, Moi A, Cistac P, Rault T, Louf R, Funtowicz M, Davison J, Shleifer S, von Platen P, Ma C, Jernite Y, Plu J, Xu C, Le Scao T, Gugger S, Drame M, Lhoest Q, Rush A (2020) Transformers: state-of-the-art natural language processing.

In: Proceedings of the 2020 conference on empirical methods in natural language processing: system demonstrations, pp 38–45. https://aclanthology.org/2020.emnlp-demos.6

Zhang Z, Guo W, Meng X, Wang Y, Wang Y, Jiang X, Liu Q, Yang Z (2023) HyperPELT: unified parameter-efficient language model tuning for both language and vision-and-language tasks. In: Findings of the association for computational linguistics: ACL 2023, Toronto, Canada, pp 11442–11453. https://aclanthology.org/2023.findings-acl.725

Chapter 5
LoDA: Low-Dimensional Adaptation of Large Language Models

Jing Liu, Toshiaki Koike-Akino, Pu Wang, Matthew Brand, Kieran Parsons, and Ye Wang

Abstract Parameter-efficient fine-tuning (PEFT) has recently garnered significant attention, due to the enormous size of LLMs. Among various PEFT methods, low-rank adaptation (LoRA) demonstrates comparable performance to full fine-tuning, despite having significantly fewer trainable parameters. In this work, we first generalize LoRA from a low-rank linear adaptation/mapping to low-dimensional, non-linear adaptation/mapping, which we name "low-dimensional adaptation" (LoDA). We also propose LoDA+, which further improves the expressiveness of the non-linear adaptation, while still using nearly the same number of tunable parameters as LoRA. Both LoDA and LoDA+ include LoRA as a special case. To improve computational efficiency at the inference phase, we further propose R-LoDA(+) and S-LoDA(+), by replacing the pre-trained weight matrix with its low-rank or sparse approximation, which is frozen during fine-tuning. Empirical evaluations on natural language generation tasks demonstrate that variants of LoDA outperform LoRA and other baselines.

5.1 Introduction

LLMs such as ChatGPT (Achiam et al. 2023), Gemini (Anil et al. 2023), LLaMA2 (Touvron et al. 2023), and Claude 3 (Anthropic 2024) have shown great promise in generating human-like text and have sparked excitement about their potential applications across various industries. The sizes of large LLMs have grown at an unprecedented rate, with current models boasting parameter counts in the hundreds of billions or even trillions, necessitating massive amounts of computational resources for training and inference. Recent studies show that the performance of a pre-trained language model (PLM) can be significantly improved by further fine-tuning on domain-specific data (Hu et al. 2022). Therefore, fine-tuning PLMs for domain-specific tasks has become the *de facto* procedure. However, full fine-tuning of such

J. Liu (✉) · T. Koike-Akino · P. Wang · M. Brand · K. Parsons · Y. Wang
Mitsubishi Electric Research Laboratories (MERL), Cambridge, MA, USA
e-mail: jiliu@merl.com

© The Author(s), under exclusive license to Springer Nature Switzerland AG 2025
P. Passban et al. (eds.), *Enhancing LLM Performance*, Machine Translation: Technologies and Applications 7, https://doi.org/10.1007/978-3-031-85747-8_5

LLMs is still very expensive. For instance, fine-tuning a 65 billion-parameter model requires more than 780 GB of memory (Dettmers et al. 2023). Parameter-efficient fine-tuning (PEFT) fine-tunes only a small set of parameters, which may be a subset of the existing model parameters or a set of newly added parameters, thereby greatly reducing computational and memory costs. Another advantage of PEFT is that, in addition to the pre-trained model, only a small number of (extra) model parameters need to be stored for each fine-tuned task. While PEFT greatly saves storage for multiple downstream tasks, full fine-tuning needs to generate a new large model for each such task.[1] Besides parameter savings, PEFT makes it possible to quickly adapt to new tasks without catastrophic forgetting (Pfeiffer et al. 2020), which has often been observed during full fine-tuning of LLMs. PEFT approaches have also been shown to be better than full fine-tuning in low-data regimes (Hu et al. 2022; Li and Liang 2021).

Unsurprisingly, therefore, many PEFT methods have been proposed. Prefix tuning (Li and Liang 2021) and prompt tuning (Lester et al. 2021) prepend some tunable prefix tokens to the input or hidden layers, and only train these soft prompts during fine-tuning. Several adapter tuning methods (Houlsby et al. 2019; Pfeiffer et al. 2020; Rebuffi et al. 2017; Rücklé et al. 2020) insert (and tune) small neural modules called adapters (see Chap. 4, Sect. 4.2 for more details) to some layers of the PLM. More recently, Hu et al. (2022) propose to use low-rank decomposition matrices to approximate the parameter update of the weight matrix of a dense layer. In particular, they propose to update the Query and Value projection matrices in the Transformer architecture, which shows promising performance and has become a popular PEFT tool for LLMs in modern libraries, e.g., Hugging Face PEFT (Mangrulkar et al. 2022). For a comprehensive review and comparison, we refer interested readers to recent surveys (Ding et al. 2023; Lialin et al. 2023; Pfeiffer et al. 2023; Sabry and Belz 2023).

LoRA is motivated by the hypothesis that the change in the model (to adapt to a related downstream task) is intrinsically low-dimensional (Aghajanyan et al. 2021). LoRA constrains the change in weights during model adaptation to be low-rank, leading to the Low-Rank Adaptation (LoRA) approach (Hu et al. 2022). For a dense layer of the PLM, its original weight parameters, e.g., $W \in \mathbb{R}^{d \times d}$ (blue parts of Fig. 5.1a, also labeled with •) is frozen. During fine-tuning, LoRA uses low-rank decomposition matrices $A \in \mathbb{R}^{d \times r}$ and $B \in \mathbb{R}^{r \times d}$ to constrain the weight update $\Delta W = AB$ (parts without • in Fig. 5.1a). As the rank r is typically set to be very small, the number of parameters in A and B are significantly less than the original W.

We view the neural network as a function, and the change in the neural network during the task adaptation can be more generally viewed as the change in its mapping. Let the input to the dense layer be denoted by x, and the output of the pre-trained

---

[1] For example, in Hu et al. (2022), a GPT-3 175B model of size 350GB is fine-tuned with a rank-4 LoRA adapter that requires only 35MB for each downstream task. Storing 100 adapted models requires only 350GB + 35MB $\times$ 100 $\approx$ 354GB as opposed to 350GB $\times$ 100 $\approx$ 35TB for full fine-tuning over 100 tasks.

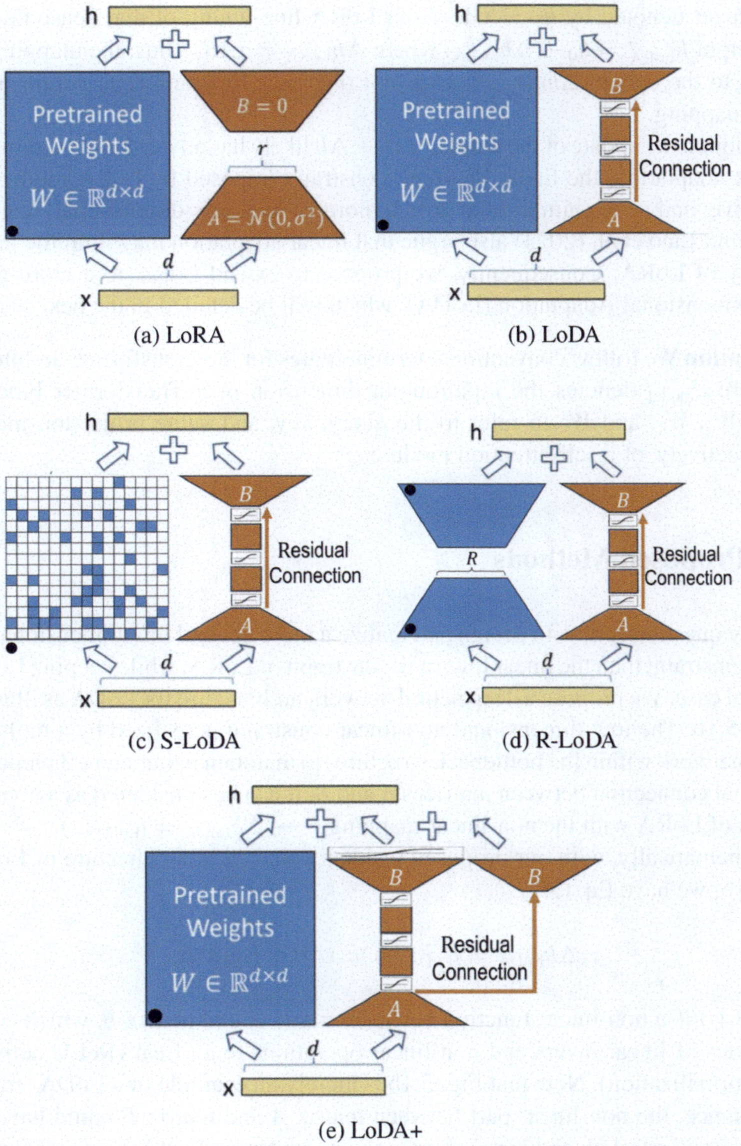

Fig. 5.1 Overview of **a** LoRA; **b** LoDA; **c** S-LoDA; **d** R-LoDA; **e** LoDA+. The blue part (labeled with •) is frozen during fine-tuning, and only other parts are trained. In LoDA+, there is essentially only one matrix B, but the non-linear part has additional non-linear operations after B (e.g., LeakyReLU)

dense layer denoted by $h_0 = xW$. After LoRA fine-tuning of that dense layer, the new output $h'_{\text{LoRA}} = h_0 + \Delta h_{\text{LoRA}}$, where $\Delta h_{\text{LoRA}} = xAB$. Thus, the mapping from input x to the update $\Delta h_{\text{LoRA}} = xAB$ by LoRA is a low-rank (i.e., r-dimensional) linear mapping.

Although the update of the mapping $x \to \Delta h$ likely has an intrinsic low dimension for task adaptation, the linear low-rank constraint imposed by LoRA might be too restrictive, and so we aim to relax it to a more general low-dimensional, non-linear constraint. Liao et al. (2023) also argue that linear adaptation may limit the learning capacity of LoRA. Consequently, we propose to extend LoRA to a more general Low-Dimensional Adaptation (LoDA), which will be detailed in the next section.

Notation We follow conventional terminologies for the Transformer architecture, where d_{model} denotes the input/output dimension of a Transformer block. We use W_{q}, W_{k}, and W_{v} to refer to the query, key, and value projection matrices, respectively, of a self-attention module.

5.2 Proposed Methods

One key question is how to design and realize a more general low-dimensional, non-linear constraint than the linear low-rank constraint of LoRA, while keeping LoRA as a special case. We propose a deep neural network architecture for LoDA as illustrated in Fig. 5.1b. The low-dimensional non-linear constraint is realized by a multi-layer neural network within the bottleneck structure (to maintain parameter efficiency) and a residual connection between matrices A and B. It can be considered as a non-linear version of LoRA with the non-linear mapping $x \to \Delta h_{\text{LoDA}} \doteq f_{\text{LoDA}}(x)$.

Mathematically, with our proposed residual connection architecture of LoDA in Fig. 5.1b, we have Eq. (5.1):

$$\Delta h_{\text{LoDA}} = f_{\text{LoDA}}(x) = xAB + f_1(xA)B, \tag{5.1}$$

where $f_1(\cdot)$ is a non-linear function between matrix A and matrix B, which consists of a series of linear layers and non-linear operations (e.g., LeakyReLU activation, layer-normalization). Note that Fig. 5.1b is merely an example of a LoDA structure. For instance, the non-linear part between matrix A and matrix B could have more layers than illustrated, and may use non-square matrices. LoRA is a special case of LoDA if $f_1(xA)B$ is zero (e.g., a hidden layer's weights in LoDA are zero) or if $f_1(\cdot)$ is linear, for example.

5.2.1 Extension to LoDA+

Although LoDA generalizes LoRA from a low-rank linear mapping/adaptation to a low-dimensional, non-linear mapping/adaptation, and keeps LoRA as a special case, the image of such a non-linear mapping still lies in a low-dimensional linear subspace (i.e., the range of matrix B). We investigate whether it is possible to further generalize that to a low-dimensional (non-linear) manifold, while keeping LoRA as a special case, and using almost the same number of tunable parameters as LoRA. We further propose the mapping in (5.2) for LoDA+, illustrated in Fig. 5.1e, that affirms the possibility:

$$\Delta h_{\text{LoDA+}} = f_{\text{LoDA+}}(x) = xAB + f_2(f_1(xA)B) \tag{5.2}$$

Note that the key difference between LoDA and LoDA+ is the additional non-linear function $f_2(\cdot)$, e.g., non-linear activation and/or layer-normalization. With this additional non-linear function, the image of the mapping $f_{\text{LoDA+}}$ becomes the combination of a linear subspace (the first term in Eq. (5.2)) and a non-linear manifold (the second term in Eq. (5.2)). For convenience, we will use LoDA(+) to represent both 'LoDA and LoDA+'.

Viewing LoDA(+) as Deep Parallel Adapters: He et al. (2022) viewed LoRA as a parallel adapter. Similarly, the proposed LoDA can be viewed as a *deep* parallel adapter. Recent work (Zhu et al. 2021; He et al. 2022) propose to use the traditional shallow adapter in a parallel fashion instead of the usual sequential fashion. The shallow adapter there only has a down-projection layer, followed by a non-linear activation function (typically ReLU), then an up-projection layer, and the adapter attaches to the input and output of the Attention module or the Feed-Forward Network module of a Transformer block in an LLM. We refer the interested readers to (Hu et al. 2023), Fig. 1) and (He et al. (2022, Table 1) for more details. In contrast, the proposed LoDA, which aims at learning a low-dimensional, non-linear mapping, has a *deep* structure to capture the underlying nonlinearity. Further, in LLMs, LoDA and LoRA are attached to W_q and W_v, but not attached to the whole Attention module nor to the Feed-Forward Network module of the Transformer block. Also, it is interesting to note that LoDA has a Residual Connection *inside*, that is between the output of matrix A and the input of matrix B (see Fig. 5.1b), which is different from the existing adapters. More interestingly, LoDA+ can be viewed as a deep+shallow dual parallel adapter, where the shallow and deep parts correspond to the first and second terms in Eq. (5.2), respectively.

5.2.2 S-LoDA(+) and R-LoDA(+)

As the dimension of the non-linear layers in LoDA(+) is restricted to a very small value r, the additional computational cost during the inference is very small (see Sect. 5.3.2 for more details). Furthermore, as LoDA(+) runs in parallel with the pre-trained weight matrix W, it will not introduce a noticeable delay in the overall inference with parallelization, unlike the sequential adapters in the literature. With LoDA(+), the main computational bottleneck is still W of the PLM. We explore the extent to which it is possible to further improve the computational efficiency of a LoDA(+) fine-tuned model, and even whether it can be significantly better than the pre-trained model.

We observe that the combined projection matrix $W_{\text{proj}} = [W_q, W_k, W_v]$ inside the Attention module of GPT2-medium (with size 1024×3072) can be well-approximated by a relatively low-rank matrix or a relatively sparse matrix. More specifically, Fig. 5.2a shows the percentage of total energy (i.e., $\sum_{i=1}^{R} \sigma_i^2 / \sum_{i=1}^{1024} \sigma_i^2$) with respect to the number of top singular values R of W_{proj} in the first Transformer block of GPT2-medium. We see that using only the top 300 singular value components of W_{proj} preserves over 93% of its total energy. Figure 5.2 shows the percentage of total energy w.r.t. the percentage of nonzero entries of W_{proj} (by zeroing out smaller magnitude weights). We see that keeping 40% of the larger magnitude entries of W_{proj} can preserve over 96% of its total energy.

The aforementioned questions and observations motivate us to further propose R-LoDA(+) and S-LoDA(+), which are LoDA(+) combined with the low-**R**ank or **S**parsified approximations of W, which are frozen during fine-tuning, while the

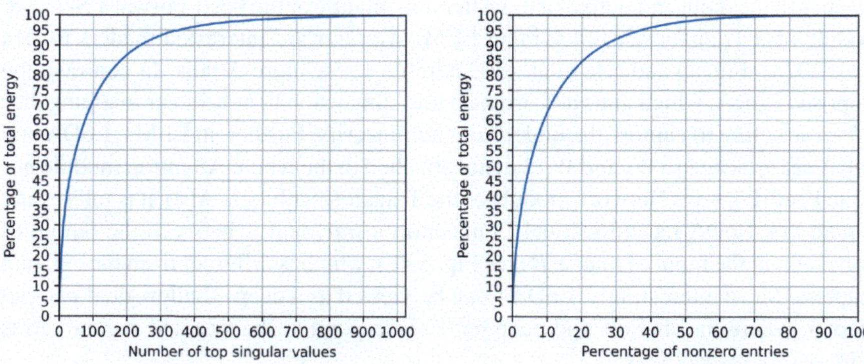

(a) (b)

Fig. 5.2 **a** Percentage of total energy w.r.t. the number of top singular values of W_{proj}. **b** Percentage of total energy w.r.t. the percentage of nonzero entries of W_{proj} (by zeroing out smaller magnitude weights) adapter is trained/fine-tuned for this approximate W. See Fig. 5.1c and d for illustrations.[2] Our empirical investigation shows that even when the pre-trained projection matrix W_{proj} is low-rank approximated or sparsified, combining with LoDA(+) can still achieve competitive performance. More importantly, the total inference cost of the R-LoDA(+) or S-LoDA(+) fine-tuned model can be significantly lower than the pre-trained model. The analysis of the computational costs can be found in Sect. 5.3.2, and Table 5.2 in Sect. 5.4 also demonstrates the significant computational savings for the GPT2-medium model.

5.3 Discussion

5.3.1 Number of Fine-Tuning Parameters

The number of trainable parameters of LoRA and the proposed methods are determined by the bottleneck dimension r and the shape of the original weights. More specifically, in Fig. 5.1, the matrices A and B across all methods have dimensions d times r. The proposed methods have two additional r by r bottleneck matrices. Their non-linear activation function is LeakyReLU with a fixed slope of 0.8, and their layer-normalization is not trainable. Accordingly, the total number of trainable parameters for LoRA is $2rdL$, and for all proposed methods is $2(rd + r^2)L$, where L is the number of weight matrices that we apply LoRA/LoDA(+)/S-LoDA(+)/R-LoDA(+) to. Note that they are almost the same when $r \ll d$. For example, in GPT2-medium, $d = d_{\text{model}} = 1024$ and $r = 4$ are used for all methods by default, so r^2 is negligible compared to rd. As in LoRA, we only apply the proposed methods to W_q and W_v (of shape $d_{\text{model}} \times d_{\text{model}}$) in the self-attention module.

5.3.2 Computational Efficiency During Inference

In Fig. 5.1, let the input embeddings be $X \in \mathbb{R}^{n \times d}$, where n is sequence length. For the LoDA(+) part, recall that $A \in \mathbb{R}^{d \times r}$, $B \in \mathbb{R}^{r \times d}$, and the two bottleneck matrices in Fig. 5.1b–e are r by r square matrices, and there are some non-linear activations and/or layer-normalization layers. The computational complexity of a LoDA(+) adapter during the inference is $O(rdn + dn + r^2n + rn)$, where the dominant part is $O(rdn)$, which is much lower than computing XW_q (or XW_v), which costs $O(d^2n)$, since $r \ll d$ (recall that in GPT2-medium, $d = d_{\text{model}} = 1024$ and $r = 4$ are used for all methods by default).

[2] If applying R-LoDA (or S-LoDA) on W_q and W_v, one could approximate W_q and W_v separately, but we directly approximate the whole $W_{\text{proj}} = [W_q, W_k, W_v]$ to make the model inference more efficient.

For R-LoDA and S-LoDA, as mentioned earlier, if applying them on W_q and W_v, one could low-rank approximate (or sparsify) W_q and W_v separately. To make the model inference more efficient, we directly approximate the whole $W_{proj} = [W_q, W_k, W_v]$ instead. More specifically, for $W_{proj} \in \mathbb{R}^{d\times 3d}$, we approximate it using the product of matrix $W_A \in \mathbb{R}^{d\times R}$ and matrix $W_B \in \mathbb{R}^{R\times 3d}$, i.e., $W_A W_B$. The computational cost for XW_{proj} is $3d^2n$ MACs (Multiply-Accumulate Operations), while the computational cost for the low-rank version $(XW_A)W_B$ is $ndR + nR3d = 4Rdn$ MACs, which is lower than the former as long as $R < 3d/4$ (e.g., in GPT2-medium, $d = 1024$, so we only need $R < 768$). One can calculate that the R-LoDA(+) fine-tuned model is computationally more efficient than the pre-trained model during inference, even setting R as high as 700 in our experimental settings.

Similarly, for S-LoDA(+), the computational cost for the sparsified version XW_{Sparse} is $s \times 3d^2n$ MACs, where s is the fraction of nonzero entries in W_{proj}. While the added computational cost of the LoDA+ adapters on W_q and W_v is $2(2rd + 2r^2 + 4r)n$ MACs, which is relatively negligible since $r \ll d$, and the total computational saving is then approximately $(1 - s) \times 3d^2n$ MACs. Reducing the computational complexity with R-LoDA(+) and S-LoDA(+) can directly decrease total power consumption in inference.

Table 5.1 GPT2-medium with different adaptation methods on E2E NLG challenge. For all metrics, higher is better

Method	Approx	# Trainable	E2E NLG challenge				
	W_{proj}	Parameters	BLEU	NIST	MET	ROUGE-L	CIDEr
FT*	No	354.92M	68.2	8.62	46.2	71.0	2.47
Adapter^{L}*	No	0.37M	66.3	8.41	45.0	69.8	2.40
Adapter^{L}*	No	11.09M	68.9	8.71	46.1	71.3	2.47
Adapter^{H}*	No	11.09M	67.3$_{\pm.6}$	8.50$_{\pm.07}$	46.0$_{\pm.2}$	70.7$_{\pm.2}$	2.44$_{\pm.01}$
PreLayer*	No	0.35M	69.7	8.81	46.1	71.4	2.49
FT^{Top2}*	No	25.19M	68.1	8.59	46.0	70.8	2.41
FT^{W_q,W_v}	No	48.00M	69.4$_{\pm.1}$	8.74$_{\pm.02}$	46.0$_{\pm.0}$	71.0$_{\pm.1}$	2.48$_{\pm.01}$
One-shot	No	–	14.6	2.86	27.5	40.3	0.82
LoRA	No	0.38M	69.0$_{\pm.7}$	8.69$_{\pm.07}$	46.5$_{\pm.2}$	71.3$_{\pm.4}$	2.51$_{\pm.00}$
LoDA	No	0.38M	**70.2**$_{\pm.3}$	**8.83**$_{\pm.03}$	**46.6**$_{\pm.1}$	**71.6**$_{\pm.1}$	**2.53**$_{\pm.01}$
S-LoDA	**Keep 40%**	0.38M	**70.2**$_{\pm.3}$	**8.83**$_{\pm.03}$	**46.6**$_{\pm.1}$	**71.6**$_{\pm.1}$	**2.53**$_{\pm.01}$
R-LoDA	**Rank300**	0.38M	**69.7**$_{\pm.2}$	8.79$_{\pm.03}$	**46.7**$_{\pm.0}$	**71.5**$_{\pm.3}$	**2.52**$_{\pm.00}$
LoDA+	No	0.38M	**69.9**$_{\pm.3}$	**8.81**$_{\pm.04}$	**46.5**$_{\pm.0}$	**71.4**$_{\pm.0}$	**2.52**$_{\pm.00}$
S-LoDA+	**Keep 40%**	0.38M	69.6$_{\pm.5}$	8.77$_{\pm.06}$	**46.7**$_{\pm.1}$	**71.6**$_{\pm.2}$	2.50$_{\pm.01}$
R-LoDA+	**Rank300**	0.38M	**70.1**$_{\pm.4}$	**8.81**$_{\pm.05}$	46.4$_{\pm.1}$	**71.6**$_{\pm.3}$	**2.52**$_{\pm.01}$

*indicates numbers published in previous work, as compiled by Hu et al. (2022)

5.4 Empirical Studies

We focus on natural language generation (NLG) tasks, and we follow the setup of Hu et al. (2022) and Li and Liang (2021) on GPT2-medium (Radford et al. 2019) for a direct comparison. We compare the downstream task performance of our proposed methods with LoRA, adapter tuning methods by Houlsby et al. (2019) ($\text{Adapter}^{\text{H}}$) and Lin et al. (2020) ($\text{Adapter}^{\text{L}}$), prefix-layer tuning (PreLayer), full fine-tuning (FT), and fine-tuning the top-2 layers (FT^{Top2}), similar to Hu et al. (2022). We also compare direct fine-tuning of the projection matrices W_{q} and W_{v} (denoted as $\text{FT}^{W_{\text{q}},W_{\text{v}}}$). As a reference, we also tested the one-shot in-context learning performance of the pre-trained GPT2-medium model, by providing the task description and one example in the prompt.

For LoDA(+), we simply set the hyperparameters (e.g., bottleneck dimension $r = 4$, learning rate, etc.) to the same as that used by LoRA (indicated in Table 11 of Hu et al. (2022)[3]) without tuning, which may favor LoRA. For R-LoDA(+) and S-LoDA(+), since the pre-trained W_{proj} is approximated, training a few more epochs may be needed to recover some details that are potentially lost during approximation. Thus, we train R-LoDA(+) and S-LoDA(+) up to 10 epochs and choose the best result from epoch 5 and epoch 10. The experiments were run on an NVIDIA A40 GPU with 48 GB memory.

We first evaluated on the E2E NLG challenge (Novikova et al. 2017) dataset, which is a dataset for training end-to-end NLG systems and is commonly used for data-to-text evaluation. The dataset consists of approximately 42K training, 4.6K validation, and 4.6K testing examples from the restaurant domain. It is released under the Creative Commons BY-NC-SA 4.0 license. Table 5.1 compares the performance of different methods on this dataset. One-shot in-context learning[4] performs much worse than other methods which fine-tune the model. LoDA, LoDA+, and S-LoDA outperform the baselines (including Fine-Tuning methods) on all 5 evaluation metrics. Other variants R-LoDA(+) and S-LoDA+ perform better than or at least on-par with LoRA and other baselines.

---

[3] We do not know the random seeds used in Hu et al. (2022). So we run LoDA(+) and LoRA with the same random seeds for fair comparisons. On DART, we cannot reproduce the results of LoRA using the default of 5 epochs, and we run 10 epochs instead to obtain results similar to that reported in Hu et al. (2022).

[4] The task description and one example that we provided in the prompt are as follows: Generate a restaurant description from the table. Here is an example: name : The Eagle | Type : coffee shop | food : Japanese | price : less than £20 | customer rating : low | area : riverside | family friendly : yes | near : Burger King. The generated description is: The Eagle is a low rated coffee shop near Burger King and the riverside that is family friendly and is less than £20 for Japanese food.

We also perform experiments on the DART dataset (Nan et al. 2021) following the setup of Hu et al. (2022) and Li and Liang (2021). This open-domain data-to-text dataset has a total of 82K examples. DART presents a significantly larger and more complex data-to-text task compared to the E2E NLG challenge dataset (Novikova et al. 2017). This dataset is released under the MIT license. We evaluate with the BLEU (Papineni et al. 2002), METEOR (Lavie and Agarwal 2007), and TER (Snover et al. 2006) metrics, similar to Hu et al. (2022), but with slightly higher precision. The evaluation code is based on the source code of LoRA (Hu et al. 2022), which is available on GitHub.[5] The results are shown in Table 5.2, which also include the number of MACs per token for computing Query/Key/Value during inference in the fine-tuned model. Note that the LoRA fine-tuned model has the same computational cost as the pre-trained model if the adapter is merged into the pre-trained weight, otherwise its computational cost is almost the same as LoDA. Also note that the number of trainable parameters in FT^{W_q, W_v} accounts for nearly 1/7 of the total model parameters, and is 126 times more than that of LoRA and proposed methods. LoDA and especially LoDA+ again outperform LoRA and FT^{W_q, W_v}, and their computational cost during inference is only slightly higher than the pre-trained model.

We notice that higher rank and less sparseness to approximate W_{proj}, respectively, in R-LoDA(+) and S-LoDA(+), are needed for DART, as it is a more complex task than the E2E NLG challenge. Nevertheless, even with pruning 40% of the entries in W_{proj}, S-LoDA(+) can outperform LoRA and FT^{W_q, W_v}. For R-LoDA(+), approximating W_{proj} with the 300 top singular value components seems insufficient on DART, while it is sufficient on E2E NLG challenge in Table 5.1 (as R-LoDA and R-LoDA+ outperform baselines on E2E NLG challenge with Rank = 300). This is likely because the E2E NLG challenge is a relatively easier downstream task, so a rough low-rank approximation of the pre-trained weights combined with LoDA(+) fine-tuning is sufficient. R-LoDA+ with Rank = 500 shows reasonable performance. We observe a trade-off between efficiency and accuracy in R-LoDA(+) and S-LoDA(+). This may shed light on how to choose the Rank and Sparsity in R-LoDA(+) and S-LoDA(+), which depends on the downstream task as well as its relation to the pre-trained tasks. Such (downstream) task-dependent auto-configuration of Rank and Sparsity is a topic for our future work.

Most notably, the number of MACs per token for computing Query/Key/Value during inference, in both the S-LoDA(+) and R-LoDA(+) fine-tuned models, is significantly lower than the pre-trained model. For example, S-LoDA+ that prunes 40% of the entries in W_{proj} can significantly reduce the number of MACs, while having slightly better performance than LoRA and FT^{W_q, W_v}.

[5] https://github.com/microsoft/LoRA.

Table 5.2 GPT2-medium with different adaptation methods on DART. For TER metric, lower is better. The number of MACs per token of computing Query/Key/Value during inference in the LoRA fine-tuned model is 75.5 million if the LoRA adapter is merged into the pre-trained weight W, otherwise it is 75.9 million

Method	Approx W_{proj}	MACs/token of compute Q/K/V	# Trainable parameters	DART		
				BLEU↑	MET↑	TER↓
FT^{W_q,W_v}	No	75.5M	48.00M	$47.1_{\pm.1}$	$\mathbf{36.0}_{\pm.0}$	$0.480_{\pm.000}$
LoRA	No	75.5/75.9M	0.38M	$47.2_{\pm.1}$	$\mathbf{36.0}_{\pm.0}$	$0.480_{\pm.000}$
LoDA	No	75.9M	0.38M	$\mathbf{47.3}_{\pm.1}$	$\mathbf{36.0}_{\pm.0}$	$0.480_{\pm.000}$
S-LoDA	**Keep 60%**	45.7M	0.38M	$\mathbf{47.3}_{\pm.2}$	$\mathbf{36.0}_{\pm.0}$	$\mathbf{0.477}_{\pm.006}$
S-LoDA	**Keep 50%**	38.1M	0.38M	$47.1_{\pm.2}$	$\mathbf{36.0}_{\pm.0}$	$0.480_{\pm.000}$
R-LoDA	**Rank500**	49.5M	0.38M	$46.8_{\pm.7}$	$35.9_{\pm.1}$	$0.483_{\pm.006}$
R-LoDA	**Rank400**	39.7M	0.38M	$46.6_{\pm.2}$	$35.9_{\pm.1}$	$0.483_{\pm.006}$
R-LoDA	**Rank300**	29.9M	0.38M	$46.5_{\pm.2}$	$35.5_{\pm.4}$	$0.487_{\pm.006}$
LoDA+	No	76.1M	0.38M	$\mathbf{47.3}_{\pm.2}$	$\mathbf{36.0}_{\pm.0}$	$\mathbf{0.477}_{\pm.006}$
S-LoDA+	**Keep 60%**	45.9M	0.38M	$\mathbf{47.3}_{\pm.1}$	$\mathbf{36.0}_{\pm.0}$	$\mathbf{0.473}_{\pm.006}$
S-LoDA+	**Keep 50%**	38.3M	0.38M	$47.1_{\pm.2}$	$\mathbf{36.0}_{\pm.0}$	$0.480_{\pm.000}$
R-LoDA+	**Rank500**	49.7M	0.38M	$47.1_{\pm.5}$	$35.9_{\pm.1}$	$0.480_{\pm.000}$
R-LoDA+	**Rank400**	39.9M	0.38M	$46.7_{\pm.2}$	$35.9_{\pm.1}$	$0.483_{\pm.006}$
R-LoDA+	**Rank300**	30.1M	0.38M	$46.3_{\pm.6}$	$35.6_{\pm.5}$	$0.483_{\pm.006}$

5.4.1 *Why LoDA(+) Outperforms FT^{W_q,W_v}*

From Tables 5.1 and 5.2, one can see that LoDA and LoDA+ consistently outperform FT^{W_q,W_v} on all evaluation metrics. Recall that LoDA(+) are applied to projection matrices W_q and W_v, while FT^{W_q,W_v} directly fine-tunes the whole matrices W_q and W_v. Naturally, one may question why LoDA(+) does better.

It is important to note that directly fine-tuning the weight matrix W of a dense layer still retains a linear mapping. Let the input to that dense layer be denoted by x, and the output of the pre-trained dense layer be denoted by $h_0 = xW$. After directly fine-tuning that dense layer, we have $W' = W + \Delta W$, and the new output $h'_{FT^W} = xW' = xW + x\Delta W = h_0 + \Delta h_{FT^W}$, where $\Delta h_{FT^W} = x\Delta W$. So the mapping from input x to the update Δh_{FT^W} is still a linear mapping, though this mapping is generally not low-rank.[6]

[6] The definition of the rank of a linear mapping can be found at https://en.wikipedia.org/wiki/Linear_map.

In contrast, the mappings of LoDA and LoDA+ in Eqs. (5.1) and (5.2) are non-linear, and cannot be expressed in the form of $\Delta h = x\Delta W$. This is in line with the observation in Eq. 5 of Liao et al. (2023), when the authors discuss the limited learning capacity of LoRA. From that perspective, our proposed LoDA(+) can be viewed as expanding the learning capacity of LoRA, which further explains why LoDA(+) performs better.

5.4.2 Effect of the Bottleneck Dimension

We further study the effect of the bottleneck dimension r of LoDA(+) adapters in GPT2-medium using the E2E NLG challenge dataset, and also include LoRA (under the same random seeds) as a baseline. The hyperparameters (e.g., learning rate) of LoDA(+) are set to be the same as that used by LoRA (indicated in Table 11 of Hu et al. (2022)) without tuning, which may favor LoRA, but the main purpose here is to study the effect of the bottleneck dimension in LoDA(+). Table 5.3 shows the performance of each adapter under different bottleneck dimensions r. LoDA+ and LoRA achieve their best performance roughly around $r = 128$, while further increasing r does not show apparent improvement. Interestingly, LoDA achieves favorable performance at bottleneck dimensions of both $r = 4$ and $r = 256$.

5.5 Conclusion and Future Work

We have generalized LoRA to the framework of LoDA(+), where LoRA is a special case, and have demonstrated their very promising performance. We also extended LoDA(+) to R-LoDA(+) and S-LoDA(+), by applying low-rank and sparse approximation, which achieves similar performance, while drastically improving computational efficiency. One future direction is to approximate W with other structured matrices, e.g., block-sparse matrix, Monarch matrix (Dao et al. 2022), or with quantization of the pre-trained model, such as in QLoRA (Dettmers et al. 2023).

Table 5.3 GPT2-medium with different adaptation methods and corresponding bottleneck dimensions r, on E2E NLG challenge. For all metrics, higher is better. For each adapter, bold fonts indicate its best metric score among tested bottleneck dimensions

Method	Bottleneck	E2E NLG challenge				
	Dimension r	BLEU	NIST	MET	ROUGE-L	CIDEr
LoRA	2	$69.4_{\pm.5}$	$8.74_{\pm.07}$	$46.5_{\pm.1}$	$71.4_{\pm.3}$	$2.49_{\pm.03}$
LoRA	4	$69.0_{\pm.7}$	$8.69_{\pm.07}$	$46.5_{\pm.2}$	$71.3_{\pm.4}$	$2.51_{\pm.00}$
LoRA	8	$69.4_{\pm.6}$	$8.74_{\pm.06}$	**46.6**$_{\pm.1}$	$71.7_{\pm.3}$	**2.52**$_{\pm.01}$
LoRA	16	$68.6_{\pm.5}$	$8.65_{\pm.06}$	$46.4_{\pm.2}$	$71.4_{\pm.3}$	$2.50_{\pm.01}$
LoRA	32	$69.4_{\pm.8}$	$8.74_{\pm.10}$	**46.6**$_{\pm.2}$	$71.6_{\pm.1}$	$2.51_{\pm.01}$
LoRA	64	$69.7_{\pm.1}$	**8.77**$_{\pm.02}$	$46.5_{\pm.1}$	**71.8**$_{\pm.1}$	$2.51_{\pm.01}$
LoRA	128	**69.8**$_{\pm.4}$	**8.77**$_{\pm.03}$	**46.6**$_{\pm.1}$	**71.8**$_{\pm.2}$	$2.51_{\pm.01}$
LoRA	256	$69.0_{\pm.4}$	$8.68_{\pm.04}$	**46.6**$_{\pm.2}$	$71.7_{\pm.2}$	$2.50_{\pm.01}$
LoRA	512	$69.4_{\pm.5}$	$8.72_{\pm.05}$	**46.6**$_{\pm.1}$	$71.6_{\pm.1}$	$2.51_{\pm.01}$
LoRA	1024	$69.4_{\pm.4}$	$8.72_{\pm.04}$	**46.6**$_{\pm.1}$	$71.6_{\pm.1}$	$2.51_{\pm.01}$
LoDA	2	$67.7_{1.0}$	$8.62_{\pm.14}$	$44.9_{\pm.5}$	$69.5_{\pm.7}$	$2.32_{\pm.04}$
LoDA	4	**70.2**$_{\pm.3}$	**8.83**$_{\pm.03}$	$46.6_{\pm.1}$	$71.6_{\pm.1}$	**2.53**$_{\pm.01}$
LoDA	8	$69.4_{\pm.2}$	$8.74_{\pm.04}$	$46.5_{\pm.2}$	$71.1_{\pm.2}$	$2.52_{\pm.01}$
LoDA	16	$69.6_{\pm.4}$	$8.77_{\pm.05}$	$46.6_{\pm.1}$	$71.4_{\pm.2}$	$2.51_{\pm.01}$
LoDA	32	$68.3_{\pm1.0}$	$8.63_{\pm.12}$	$46.3_{\pm.2}$	$71.0_{\pm.3}$	$2.49_{\pm.03}$
LoDA	64	$68.2_{\pm.7}$	$8.62_{\pm.09}$	$46.2_{\pm.2}$	$70.8_{\pm.4}$	$2.49_{\pm.02}$
LoDA	128	$69.9_{\pm.0}$	$8.81\ _{\pm.02}$	$46.6_{\pm.1}$	$71.6_{\pm.2}$	**2.53**$_{\pm.01}$
LoDA	256	**70.2**$_{\pm.8}$	$8.82_{\pm.09}$	**46.8**$_{\pm.1}$	**71.8**$_{\pm.3}$	**2.53**$_{\pm.02}$
LoDA	512	$68.9_{\pm.8}$	$8.69_{\pm.08}$	$46.4_{\pm.4}$	$71.5_{\pm.5}$	$2.50_{\pm.02}$
LoDA	1024	$68.9_{\pm.3}$	$8.70_{\pm.04}$	$46.5_{\pm.2}$	$71.5_{\pm.2}$	$2.51_{\pm.01}$
LoDA+	2	$66.9_{\pm.9}$	$8.52_{\pm.16}$	$44.7\ _{\pm.1}$	$69.3_{\pm.1}$	$2.35_{\pm.06}$
LoDA+	4	$69.9_{\pm.3}$	$8.81_{\pm.04}$	$46.5_{\pm.0}$	$71.4_{\pm.0}$	$2.52_{\pm.00}$
LoDA+	8	$70.0_{\pm.6}$	$8.82_{\pm.06}$	$46.6_{\pm.1}$	$71.4_{\pm.3}$	**2.54**$_{\pm.02}$
LoDA+	16	$69.6_{\pm.3}$	$8.77_{\pm.05}$	$46.6_{\pm.1}$	$71.4_{\pm.3}$	$2.52_{\pm.01}$
LoDA+	32	$69.9_{\pm.3}$	$8.79_{\pm.04}$	$46.7_{\pm.1}$	$71.7_{\pm.0}$	$2.53_{\pm.01}$
LoDA+	64	$69.9_{\pm.6}$	$8.81_{\pm.06}$	$46.7_{\pm.1}$	$71.6_{\pm.5}$	$2.52_{\pm.02}$
LoDA+	128	**70.2**$_{\pm.6}$	**8.83**$_{\pm.06}$	**46.8**$_{\pm.1}$	$71.9_{\pm.2}$	$2.53_{\pm.01}$
LoDA+	256	$69.9_{\pm.4}$	$8.79_{\pm.03}$	$46.6_{\pm.1}$	$71.7_{\pm.4}$	$2.52_{\pm.01}$
LoDA+	512	$70.1_{\pm.5}$	$8.81_{\pm.06}$	**46.8**$_{\pm.1}$	**72.0**$_{\pm.2}$	$2.53_{\pm.02}$
LoDA+	1024	$69.6_{\pm.8}$	$8.77_{\pm.11}$	$46.7_{\pm.1}$	$71.7_{\pm.2}$	$2.52_{\pm.03}$

References

Achiam J, Adler S, Agarwal S, Ahmad L, Akkaya I, Aleman FL, Almeida D, Altenschmidt J, Altman S, Anadkat S et al (2023) Gpt-4 technical report. arXiv preprint arXiv:2303.08774

Aghajanyan A, Gupta S, Zettlemoyer L (2021) Intrinsic dimensionality explains the effectiveness of language model fine-tuning. In: Proceedings of the 59th annual meeting of the association for computational linguistics and the 11th international joint conference on natural language processing (Volume 1: Long papers), Online, pp 7319–7328. https://aclanthology.org/2021.acl-long.568

Anil R, Borgeaud S, Wu Y, Alayrac JB, Yu J, Soricut R, Schalkwyk J, Dai AM, Hauth A et al (2023) Gemini: a family of highly capable multimodal models. arXiv preprint arXiv:2312.11805

Anthropic (2024) The claude 3 model family: opus, sonnet, haiku. Claude 3 technical report https://www-cdn.anthropic.com/de8ba9b01c9ab7cbabf5c33b80b7bbc618857627/Model_Card_Claude_3.pdf

Dao T, Chen B, Sohoni NS, Desai A, Poli M, Grogan J, Liu A, Rao A, Rudra A, Ré C (2022) Monarch: expressive structured matrices for efficient and accurate training. In: International conference on machine learning, PMLR, pp 4690–4721

Dettmers T, Pagnoni A, Holtzman A, Zettlemoyer L (2023) Qlora: efficient finetuning of quantized llms. arXiv preprint arXiv:2305.14314

Ding N, Qin Y, Yang G, Wei F, Yang Z, Su Y, Hu S, Chen Y, Chan CM, Chen W et al (2023) Parameter-efficient fine-tuning of large-scale pre-trained language models. Nature Mach Intell 5(3):220–235

He J, Zhou C, Ma X, Berg-Kirkpatrick T, Neubig G (2022) Towards a unified view of parameter-efficient transfer learning. In: International conference on learning representations (ICLR)

Houlsby N, Giurgiu A, Jastrzebski S, Morrone B, De Laroussilhe Q, Gesmundo A, Attariyan M, Gelly S (2019) Parameter-efficient transfer learning for nlp. In: International conference on machine learning, PMLR, pp 2790–2799

Hu EJ, yelong shen, Wallis P, Allen-Zhu Z, Li Y, Wang S, Wang L, Chen W (2022) LoRA: low-rank adaptation of large language models. In: International conference on learning representations (ICLR). https://openreview.net/forum?id=nZeVKeeFYf9

Hu Z, Wang L, Lan Y, Xu W, Lim EP, Bing L, Xu X, Poria S, Lee R (2023) LLM-adapters: An adapter family for parameter-efficient fine-tuning of large language models. In: Proceedings of the 2023 conference on empirical methods in natural language processing, association for computational linguistics, Singapore, pp 5254–5276. https://aclanthology.org/2023.emnlp-main.319

Lavie A, Agarwal A (2007) METEOR: An automatic metric for MT evaluation with high levels of correlation with human judgments. In: Proceedings of the second workshop on statistical machine translation, Prague, Czech Republic, pp 228–231. https://aclanthology.org/W07-0734

Lester B, Al-Rfou R, Constant N (2021) The power of scale for parameter-efficient prompt tuning. In: Conference on empirical methods in natural language processing. https://api.semanticscholar.org/CorpusID:233296808

Li XL, Liang P (2021) Prefix-tuning: Optimizing continuous prompts for generation. In: Proceedings of the 59th annual meeting of the association for computational linguistics and the 11th international joint conference on natural language processing (Volume 1: Long papers). Association for Computational Linguistics, pp 4582–4597. https://aclanthology.org/2021.acl-long.353

Lialin V, Deshpande V, Rumshisky A (2023) Scaling down to scale up: a guide to parameter-efficient fine-tuning. arXiv preprint arXiv:2303.15647

Liao B, Meng Y, Monz C (2023) Parameter-efficient fine-tuning without introducing new latency. ArXiv arXiv:abs/2305.16742. https://api.semanticscholar.org/CorpusID:258947572

Lin Z, Madotto A, Fung P (2020) Exploring versatile generative language model via parameter-efficient transfer learning. In: Findings of the association for computational linguistics: EMNLP 2020, pp 441–459. https://aclanthology.org/2020.findings-emnlp.41

Mangrulkar S, Gugger S, Debut L, Belkada Y, Paul S, Bossan B (2022) Peft: state-of-the-art parameter-efficient fine-tuning methods. https://www.github.com/huggingface/peft

Nan L, Radev D, Zhang R, Rau A, Sivaprasad A, Hsieh C, Tang X, Vyas A, Verma N, Krishna P, Liu Y, Irwanto N, Pan J, Rahman F, Zaidi A, Mutuma M, Tarabar Y, Gupta A, Yu T, Tan YC, Lin XV, Xiong C, Socher R, Rajani NF (2021) DART: open-domain structured data record to text generation. In: Proceedings of the 2021 conference of the North American chapter of the association for computational linguistics: human language technologies, pp 432–447. https://aclanthology.org/2021.naacl-main.37

Novikova J, Dušek O, Rieser V (2017) The E2E dataset: new challenges for end-to-end generation. In: Proceedings of the 18th annual SIGdial meeting on discourse and dialogue. Association for Computational Linguistics, Saarbrücken, Germany, pp 201–206. https://aclanthology.org/W17-5525

Papineni K, Roukos S, Ward T, Zhu WJ (2002) BLEU: a method for automatic evaluation of machine translation. In: Proceedings of the 40th annual meeting of the association for computational linguistics, Philadelphia, Pennsylvania, USA, pp 311–318. https://aclanthology.org/P02-1040

Pfeiffer J, Kamath A, Rücklé A, Cho K, Gurevych I (2020) Adapterfusion: non-destructive task composition for transfer learning. ArXiv arXiv:abs/2005.00247. https://api.semanticscholar.org/CorpusID:218470208

Pfeiffer J, Ruder S, Vulić I, Ponti EM (2023) Modular deep learning. arXiv preprint arXiv:2302.11529

Radford A, Wu J, Child R, Luan D, Amodei D, Sutskever I et al (2019) Language models are unsupervised multitask learners. OpenAI Blog 1(8):9

Rebuffi SA, Bilen H, Vedaldi A (2017) Learning multiple visual domains with residual adapters. In: Proceedings of the 31st international conference on neural information processing systems. Curran Associates Inc, Red Hook, NY, USA, p 506–516

Rücklé A, Geigle G, Glockner M, Beck T, Pfeiffer J, Reimers N, Gurevych I (2020) Adapterdrop: on the efficiency of adapters in transformers. arXiv arXiv:2010.11918

Sabry M, Belz A (2023) Peft-ref: a modular reference architecture and typology for parameter-efficient finetuning techniques. arXiv preprint arXiv:2304.12410

Snover M, Dorr B, Schwartz R, Micciulla L, Makhoul J (2006) A study of translation edit rate with targeted human annotation. In: Proceedings of the 7th conference of the association for machine translation in the Americas: technical papers, pp 223–231

Touvron H, Martin L, Stone K, Albert P, Almahairi A, Babaei Y, Bashlykov N, Batra S, Bhargava P, Bhosale S et al (2023) Llama 2: open foundation and fine-tuned chat models. arXiv preprint arXiv:2307.09288

Zhu Y, Feng J, Zhao C, Wang M, Li L (2021) Serial or parallel? plug-able adapter for multilingual machine translation 6(3). arXiv preprint arXiv:2104.08154

Chapter 6
Sparse Fine-Tuning for Inference Acceleration of Large Language Models

Eldar Kurtic, Denis Kuznedelev, Elias Frantar, Michael Goinv, Shubhra Pandit, Abhinav Agarwalla, Tuan Nguyen, Alexandre Marques, Mark Kurtz, and Dan Alistarh

Abstract We investigate the problem of accurate *sparse fine-tuning* of large language models (LLMs), that is, fine-tuning pre-trained LLMs on specialized tasks, while inducing sparsity in their weights. Our work is motivated by experiments showing that standard loss-based fine-tuning methods are not able to achieve high accuracy in this setting, especially at high sparsity targets. To address this issue, we perform a detailed study of knowledge distillation losses for fine-tuning of sparse models. We determine an L2-based distillation approach that we term 'SquareHead', which enables accurate recovery even at higher sparsities. Investigating the question of efficient inference, we show that sparse LLMs can be executed faster by taking advantage of sparsity. Specifically, we exhibit end-to-end results showing speedups enabled by sparsity, while recovering accuracy, on the following models and tasks, respectively: T5 for language translation, Whisper for speech translation, and open GPT-type models such as the Mosaic Pre-Trained Transformer (MPT) and Llama-2 models for text generation. In particular, for popular generative tasks, we show for the first time that sparse fine-tuning can reach 75% sparsity without drops in accuracy, and provide notable end-to-end speedups for inference on CPUs. Moreover, we also highlight that sparsity is compatible with other compression approaches, such as quantization.

E. Kurtic (✉) · E. Frantar · D. Alistarh
Institute of Science and Technology Austria (ISTA), Klosterneuburg, Austria
e-mail: eldar.kurtic@ist.ac.at

E. Kurtic · M. Goinv · S. Pandit · A. Agarwalla · T. Nguyen · A. Marques · M. Kurtz · D. Alistarh
Neural Magic, Somerville, USA

D. Kuznedelev
Skoltech, Moscow, Russia

Yandex, Moscow, Russia

© The Author(s), under exclusive license to Springer Nature Switzerland AG 2025
P. Passban et al. (eds.), *Enhancing LLM Performance*, Machine Translation: Technologies and Applications 7, https://doi.org/10.1007/978-3-031-85747-8_6

6.1 Introduction

Large Transformer models (Vaswani et al. 2017) have gained high popularity and adoption due to their breakthrough performance across a wide range of challenging tasks. To address their high runtime costs, several acceleration techniques have been developed (Dao et al. 2022; Dettmers et al. 2022b; Frantar et al. 2022; Dettmers and Zettlemoyer 2023). For inference acceleration, one of the most popular techniques is *quantization* (Dettmers et al. 2022a; Frantar et al. 2022; Yao et al. 2022; Dettmers and Zettlemoyer 2023). Specifically, it has been shown that LLMs can be quantized down to 4 bits per weight with negligible accuracy loss and that this can be leveraged for inference speedups (Frantar et al. 2022; Dettmers and Zettlemoyer 2023). However, quantization methods are reaching accuracy limits at around 3 bits per weight (Chee et al. 2024; Dettmers et al. 2024).

A key compression alternative to quantization is *weight sparsity* (LeCun et al. 1989), which consists of setting individual LLM connections to zero. For smaller models such as BERT (Devlin et al. 2019), it is known (Sanh et al. 2020; Kurtic et al. 2022) that high levels of sparsity can be applied during *fine-tuning*, i.e., the process by which a pre-trained model is adapted to a "downstream" task, such as question answering or text classification. However, it is not known whether similar techniques an be applied at the scale of LLMs.

In this chapter, we study sparse fine-tuning for LLMs across three modern applications: speech transcription using Whisper (Radford et al. 2023), specialized to a specific language, machine translation using T5 (Wei et al. 2021), specialized to a specific language pair (Macháček and Bojar 2014), and higher-level reasoning using the openly-available MPT (Team et al. 2023) and Llama-2 (Touvron et al. 2023) models, specialized on the grade-school math (GSM8k) task (Cobbe et al. 2021).

Accordingly, the contributions of this chapter are as follows:

- We observe that naive sparse fine-tuning (Sanh et al. 2020), which follows dense fine-tuning while gradually imposing sparsity, is challenging to apply for LLMs due to training instability. More precisely, this instability manifests itself in several forms: (1) *sudden loss increases (spikes)* at higher sparsities which lead to divergence, (2) *poor accuracy recovery* for the sparse model, as the small amount of fine-tuning data available for the specific task may not be sufficient to recover accuracy; or (3) *overfitting*, as iterating multiple times over the limited fine-tuning data leads to low *training* loss, but high *validation* loss.
- To address this, we investigate fine-tuning models sparsified via the state-of-the-art SparseGPT method (Frantar and Alistarh 2023) using various losses which incorporate standard cross-entropy and output knowledge distillation (Hinton et al. 2015). More specifically, we also investigate a type of L2-based knowledge distillation inspired by Sun et al. (2019), Frantar and Alistarh (2022), Kurtić et al. (2024) which we call *SquareHead*. We show that SquareHead distillation consistently recovers accuracy, even at high sparsities.

- On the practical side, we show that the resulting sparse models can be executed faster. For CPU inference, we leverage the DeepSparse inference engine (NeuralMagic 2021) to obtain faster execution (speedups) across all three applications. Finally, we are able to obtain remarkable CPU speedups by *jointly* leveraging sparsity and quantization. For example, we are able to achieve a 7.5x speedup with a minimal accuracy loss relative to the full precision dense baseline.

6.2 Methodology

6.2.1 Sparse Fine-tuning

Sparsification. To obtain a set of sparse models satisfying some target compression requirements, we gradually increase the sparsity level, while fine-tuning the model on the task of interest. Unless otherwise stated, we start from a list of desired sparsity levels, in increasing order, and iteratively prune and fine-tune the model while following the original fine-tuning recipe for the dense model.

Distillation strategies. LLMs are notoriously difficult to train and fine-tune (Team et al. 2023), and we found this to be particularly the case when sparsity is imposed during an often short fine-tuning cycle. Choosing the 'right' loss function is critical to obtain stable fine-tuning. Thus, to address this, we investigate knowledge distillation (KD) approaches. Specifically, a sparse "student" model is trained to mimic the output of a dense and accurate "teacher" model, which has already been fine-tuned on the target task. In this context, the most common KD strategy adds a loss term measuring the Kullback–Leibler (KL) divergence between student and teacher outputs (Hinton et al. 2015). However, we observe that one obtains better results by going further and distilling intermediate representations. We describe this process formally below.

Standard output distillation uses KL-divergence between student and teacher logits as its loss, as shown in Eq. (6.1):

$$\mathcal{L}_{\text{logit}} = D_{\text{KL}}(\theta_t || \theta_s) = \frac{1}{\sum_{i}^{B \times seq} \mathbb{1}[i \notin \mathbf{P}]} \sum_{i}^{B \times seq} \mathbb{1}[i \notin \mathbf{P}] p_{\theta_t}(\mathbf{x}_i) \log \frac{p_{\theta_t}(\mathbf{x}_i)}{p_{\theta_s}(\mathbf{x}_i)}. \tag{6.1}$$

where B stands for batch size, *seq* for sequence length, $p_{\theta_t}(\mathbf{x}_i)$ and $p_{\theta_s}(\mathbf{x}_i)$ denote output probabilities for teacher and student model respectively. The notation $\mathbb{1}[i \notin \mathbf{P}]$ means that the loss for padding tokens $\mathbf{P}$ is discarded.

To *transfer intermediate representations*, we examine normalized mean squared error (MSE) loss on each feature representation, which we call **SquareHead**, as shown in Eq. (6.2):

$$\mathcal{L}^l_{\text{feat}} = \frac{\text{MSE}(f^l_t, f^l_s)}{\text{MSE}(f^l_t, 0)}, \quad \text{shape}(f^l_s) = \text{shape}(f^l_t) = B \times seq \times d_{\text{model}}. \tag{6.2}$$

Here, f_t^l and f_s^l denote the feature map of the l-th layer of the teacher and student model, and MSE represents the mean squared error calculated as $\text{MSE}(X, Y) = \frac{1}{N}\sum_{i=0}^{N}(x_i - y_i)^2$, for N-dimensional vectors X and Y. We discard terms that correspond to padding tokens. The motivation for the normalization is that the magnitude of activations may greatly vary between different layers, thus focusing the optimization procedure to optimize the layers with the largest norm. We observed that for some models MSE loss without normalization leads to instabilities in training. The total feature loss is the sum of per-layer losses over all encoder/decoder blocks.

SquareHead Distillation. The overall SquareHead distillation loss is the sum of the original task loss and SquareHead term with equal weights: $\mathcal{L} = \mathcal{L}_{\text{task}} + \mathcal{L}_{\text{feat}}$.

As stated previously, variants of SquareHead-type losses have been used by Sun et al. (2019), Frantar and Alistarh (2022), and Kurtić et al. (2024) in the context of compressing smaller-scale models during the fine-tuning process, such as BERT-style models (Devlin et al. 2019) for question-answering or sentiment classification fine-tuning tasks. More broadly, in the context of BERT fine-tuning, it is known that variants of knowledge distillation can help reduce accuracy loss during fine-tuning (Sanh et al. 2020; Frantar and Alistarh 2022; Xia et al. 2022; Kurtic et al. 2022; Kurtić et al. 2024). Relative to this line of work, in this chapter we determine a type of distillation loss that is consistently effective for accurate sparse fine-tuning of LLMs, which has not yet been investigated.

6.2.2 *Quantization*

Besides sparsification, we apply 8-bit quantization to both weights and activations to further improve performance. Quantizing activations of LLMs is known to be challenging due to the presence of outliers (Dettmers and Zettlemoyer 2023). The most effective approach for mitigating outliers is re-scaling the activations and weights simultaneously such that the outcome of linear operators remains unaltered, but that activations are more uniform. Specifically, this makes activations easier to quantize at the cost of making weights harder to quantize. Here, we refer to this approach as *smoothing*. Smoothing for LLMs was introduced in SmoothQuant (Yao et al. 2022), but is also present in other methods such as OmniQuant (Shao et al. 2023) and Logarithmic Activation Equalization (Li et al. 2023).

However, we observe experimentally that smoothing alone is insufficient to quantize certain LLMs without significant loss in accuracy. We conclude that, at the current state of the art, one needs to skip quantization of linear operators severely affected by outliers in order to preserve accuracy. Fortunately, we find skipping quantization of only a small number of operators is enough to recover accuracy. A typical example is skipping 3 to 10 operators out of 320 in Llama-2 7B (Touvron et al. 2023). Moreover, not quantizing such operators has a negligible impact on the overall inference speed.

The best criterion to identify which layers to skip during quantization is to quantize each operator independently, one at a time, and evaluate the accuracy sensitivity due to quantization. Clearly, this strategy is costly and does not scale to larger models, but we use it in our investigations to correlate with other criteria that are simpler and cheaper to evaluate. Out of these criteria, kurtosis[1] is the one that produced the most robust results across different models. We also experimented with the range or maximum absolute value, but both approaches failed to identify outliers in some models.

We quantize all linear operators (with the exception of those skipped due to accuracy concerns, as discussed above), including operators that do not contain weights, such as attention scores and attention output. As such, the cached keys and value tensors can be stored in 8-bits, improving inference speed. Finally, we apply weight quantization in one-shot using GPTQ (Frantar et al. 2022) at the end of sparse fine-tuning.

6.3 Compression of T5 and Whisper Translation Models

In our first experimental scenario, we consider the compression of Transformer models for language translation (T5) and speech transcription (Whisper).

6.3.1 Accuracy Versus Sparsity

Experimental Setup. We consider sparsity levels corresponding to 2x, 3x, 4x, 5x, 6x, 7x, 8x, and 10x compression ratios, applied uniformly per layer with SparseGPT (Frantar and Alistarh 2023). We compare sparse fine-tuning with three variants of losses: cross entropy (original loss, $\mathcal{L}_{\text{task}}$), Standard KD ($\mathcal{L}_{\text{task}}+\mathcal{L}_{\text{logit}}$), and SquareHead KD ($\mathcal{L}_{\text{task}} + \mathcal{L}_{\text{feat}}$). The computational speedups, presented relative to the uncompressed baseline, are reported for end-to-end execution in the DeepSparse inference engine (NeuralMagic 2021) on an Intel Sapphire Rapids CPU with 8 cores (AWS m7i.4xlarge).

Compression for language translation using T5. We begin by investigating the sparsification of a pre-trained T5 model (Raffel et al. 2020), fine-tuned on the popular English-German subset of WMT14 (Bojar et al. 2014). Following standard practice, we compute BLEU scores (Papineni et al. 2002) on a validation set as a measure of the model's accuracy. Figure 6.1 shows the accuracy-vs-sparsity trade-off for various loss functions. Table 6.1 shows scores for the best-performing loss variant (SquareHead KD), together with speedups relative to the uncompressed model. For reference, the dense baseline needs 370 milliseconds (ms) to encode/decode 128 tokens. We observed that sparse training with CE and Standard KD is unstable at high sparsities; only SquareHead KD is able to produce highly sparse models with reasonable accuracy recoveries.

---

[1] Kurtosis is a statistical measure to understand the shape and characteristics of data distributions. It provides useful information about tails behaviour.

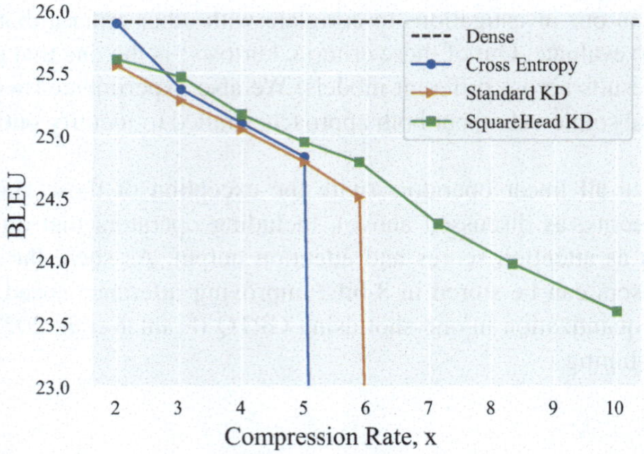

Fig. 6.1 BLEU (↑) score for variants of loss functions on English-German WMT14 and T5-Small at various compression rates

Table 6.1 BLEU score and speedups of T5-Small for various sparsities and SquareHead KD loss on English-German WMT14

Model	Sparsity (%)	BLEU (↑)	Speedup (×)
T5	0	25.9	1.0
	50	25.4	1.9
	67	25.0	2.1
	75	24.7	2.1
	80	24.6	2.2
	86	24.1	2.3
	90	23.1	2.4

Compression of speech-to-text using Whisper. In Fig. 6.2 and Table 6.2, we study compression of Whisper-Small (244M) for automatic speech recognition (ASR) on the Hindi (Hi) language subset of CommonVoice 11.0 (Ardila et al. 2020). We report word error rate (WER) (Levenshtein 1966) as a standard measure of ASR performance. For reference in terms of absolute numbers, the dense CPU baseline takes 882 ms to transcribe an audio sequence of 15 seconds. A detailed breakdown is presented in Sect. 6.3.2.

Hyperparameters. For both models, we use weight decay of 10^{-4} and a linearly decaying learning rate (LR) scheduler. The T5 model is fine-tuned with a batch-size of 128 over three epochs with 300 LR-warmup steps and peak-LR value of 2×10^{-3}. The Whisper model is fine-tuned with a batch-size of 32 over six epochs with 50 LR-warmup steps and peak-LR value of 2×10^{-4}.

Further Analysis. The results observed so far show that the SquareHead KD loss improves upon standard approaches in these two fine-tuning scenarios, especially at higher sparsities, where the latter tend to diverge. A possible explanation for the success of SquareHead could be the fact that fine-tuning with task loss only,

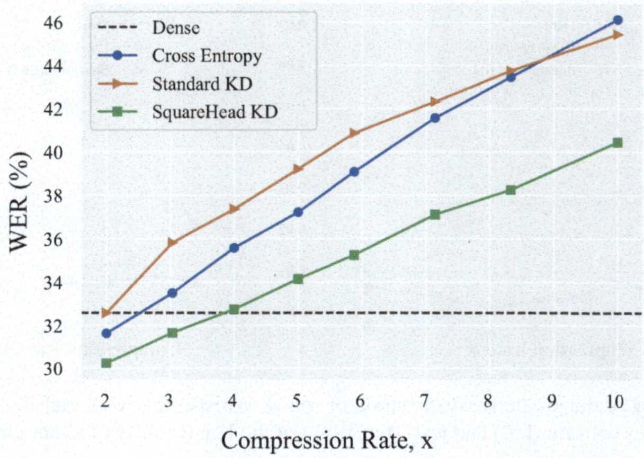

Fig. 6.2 WER (↓) for variants of loss functions on the Hindi dataset and Whisper-Small at various compression rates

Table 6.2 WER and speedups of Whisper-Small for various sparsities and SquareHead KD loss on the Hindi dataset

Model	Sparsity (%)	WER (↓)	Speedup (×)
Whisper	0	32.6	1.0
	50	30.9	1.6
	67	31.8	2.1
	75	33.1	2.2
	80	34.7	2.3
	86	37.3	2.5
	90	40.6	2.7

especially on limited data, leads to overfitting. We measured the entropy of the predictive distribution of sparse Whisper models and observed that models fine-tuned without KD have a very low entropy, i.e., make overconfident predictions, whereas SquareHead induces regularization. This behaviour is visualized in Fig. 6.3.

6.3.2 *Performance Breakdown*

Next, we would like to understand the individual performance of the encoder and decoder components in sequence-to-sequence models like T5 and Whisper. These components handle different tasks; the encoder processes input data into a context-rich representation, and the decoder translates this representation into generated

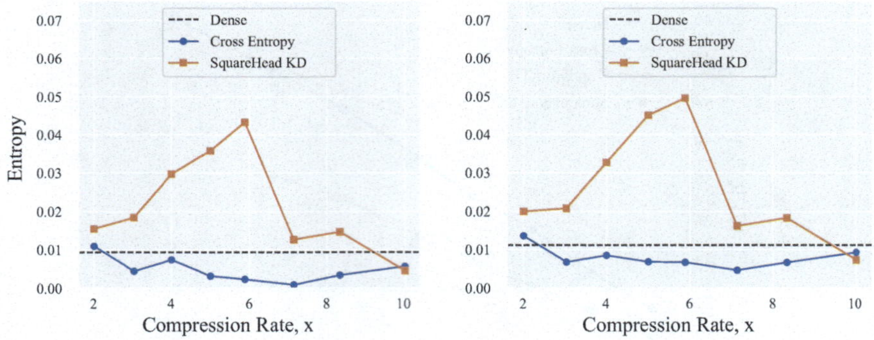

Fig. 6.3 Entropy of the predictive distribution of sparse Whisper-244M models for different fine-tuning approaches on train (left) and test (right) split of the Hindi subset of Common Voice dataset

tokens, using both the encoder's representation and its own context to produce subsequent output. While individual component performance provides insight into bottlenecks, it is also essential to evaluate end-to-end performance as it offers a more holistic understanding of how compression affects real-world application scenarios. In Figs. 6.4 and 6.5, and Tables 6.3 and 6.4 we present detailed performance breakdown from the DeepSparse engine.

Further Analysis. First, we notice that, due to similar model structures, the sparsity-related speedups are similar between the T5 and the Whisper models. (However, the latencies for Whisper are much higher.) Furthermore, note that the more compute-bound Decoder component benefits significantly more from sparsity

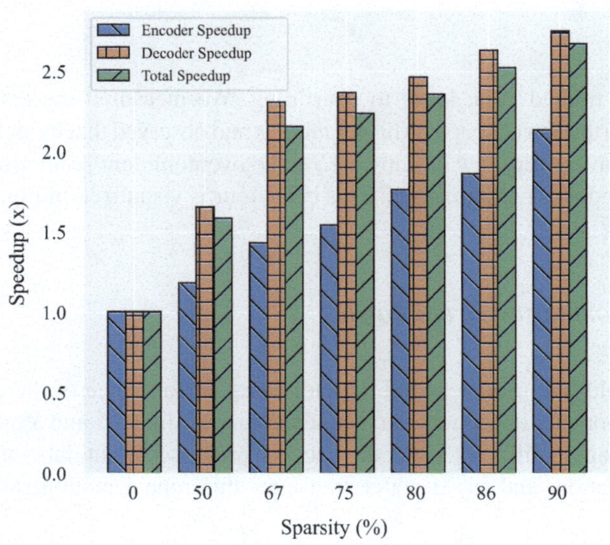

Fig. 6.4 Per-component speedups of Whisper-Small for various sparsities

Table 6.3 Encoder, decoder, and total timings of Whisper-Small for various sparsities. The total workload is generating 60 tokens from 15 seconds of audio

	Encoder		Decoder		Total	
Sparsity (%)	Speedup	Latency (ms)	Speedup	Latency (ms)	Speedup	Latency (ms)
0	1.0x	89.4	1.0x	12.8	1.0x	882.7
50	1.2x	75.5	1.6x	7.8	1.6x	558.2
67	1.4x	62.5	2.3x	5.6	2.1x	410.1
75	1.5x	58.0	2.4x	5.4	2.2x	395.5
80	1.7x	50.9	2.5x	5.2	2.3x	374.9
86	1.9x	48.0	2.6x	4.9	2.5x	350.2
90	2.1x	42.0	2.8x	4.7	2.7x	330.4

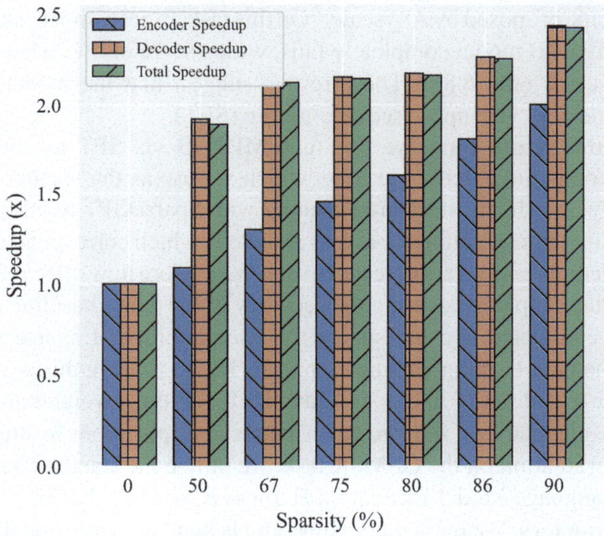

Fig. 6.5 Component speedups of T5-Small for various sparsities

Table 6.4 Encoder, decoder, and total timings of T5-Small for various sparsities. The total workload is encoding 128 tokens and generating 128 tokens

	Encoder		Decoder		Total	
Sparsity (%)	Speedup	Latency (ms)	Speedup	Latency (ms)	Speedup	Latency (ms)
0	1.0x	5.2	1.0x	3.0	1.0x	391.9
50	1.1x	4.8	1.9x	1.6	1.9x	206.9
67	1.3x	3.9	2.1x	1.4	2.1x	184.8
75	1.5x	3.6	2.1x	1.4	2.1x	183.5
80	1.6x	3.2	2.2x	1.4	2.2x	181.2
86	1.8x	2.9	2.3x	1.3	2.3x	174.2
90	2.0x	2.6	2.4x	1.2	2.4x	161.8

in terms of speedups relative to dense: for instance, on T5, Decoder speedups are between 1.6x (at 50% sparsity) and 2.7x (at 90% sparsity), whereas Encoder speedups are much more limited. Thus, the significant end-to-end speedups are mostly driven by the speedups in the Decoder.

6.4 Compression of Generative Pre-Trained Transformer Models

We investigate compression of generative models, specifically MPT-7B (Team et al. 2023) and Llama-2-7B (Touvron et al. 2023). We focus on the GSM8K task (Cobbe et al. 2021), a dataset with high quality and diverse grade school math problems, following a recipe proposed by Anyscale.[2] On this task, in zero-shot evaluation mode, the baseline MPT-7B model completely fails, with a score of 0%, whereas in 8-shot evaluation it scores only 6.8%. These results suggest that the model necessitates additional refinement via supervised fine-tuning (SFT).

Experimental setup. First, we fine-tune MPT-7B via SFT to obtain a highly accurate and competitive dense baseline, which we use as the teacher in KD runs. Then, we apply one-shot unstructured pruning with SparseGPT to 50%, 60%, 70%, and 80% sparsity targets, uniformly across all layers, which correspond to 2.0x, 2.5x, 3.3x, and 5.0x compression ratios, respectively. We explore how different sparse fine-tuning techniques help in recovering the accuracy of the dense baseline model. After fine-tuning is completed, we investigate the compatibility of sparse models with quantization via post-training quantization to INT8. To achieve this we leverage the SparseML (Kurtz et al. 2020) library and quantize to 8-bits the weights and activations of all linear weight matrices, and two batch matrix multiplications in attention layers. For accuracy evaluation on the GSM8K task, we utilize the standardized evaluation protocol via Language Model Evaluation Harness (Gao et al. 2023).

Hyperparameters. We untie input embeddings and language modelling head for compatibility with quantization via SparseML, where we quantize weights of input embeddings, and weights and activations of the language modelling head. For all sparsities, we one-shot prune the model with the default SparseGPT (Frantar and Alistarh 2023) parameters and then fine-tune for either 2 (50% and 60% sparsity) or 4 epochs (70% and 80% sparsity). We use a linearly decaying learning rate (LR) with a warmup of 20 steps, a batch-size of 32, and a sweep over 3e-5, 5e-5, 8e-5, and 1e-4 peak-LR values.

Discussion. The accuracy results for different losses are shown in Fig. 6.6. They exhibit very similar trends to the prior two applications, showing that SquareHead KD is superior to both standard CE loss and standard KD for sparse fine-tuning. The fact that the distilled models can outperform the dense baseline at low sparsity can be explained due to the effect of distillation being applied, but also possibly

[2] https://www.anyscale.com/blog/fine-tuning-llms-lora-or-full-parameter-an-in-depth-analysis-with-llama-2.

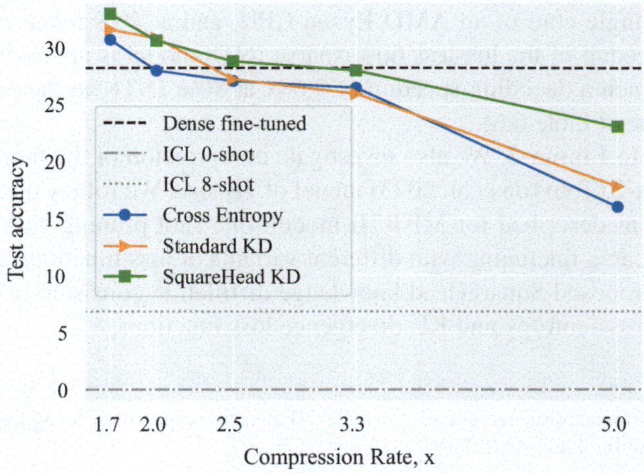

Fig. 6.6 Test accuracy for variants of loss functions on the GSM8K dataset and MPT-7B model at various compression rates. ICL stands for in-context learning results using the pre-trained dense model, without fine-tuning

Table 6.5 Test accuracy of pruned (FP32) and pruned-quantized (INT8) MPT-7B models on the GSM8K dataset. Speedups are measured relative to FP32, for end-to-end decode latency in DeepSparse (NeuralMagic 2021) using an 8-core CPU at sequence length 512

	FP32		INT8	
Sparsity (%)	Test accuracy	CPU speedup	Test accuracy	CPU speedup
0	28.2	1.0x	27.8	4.0x
40	32.9	1.5x	30.3	5.3x
50	30.6	1.8x	30.7	5.7x
60	28.8	2.1x	28.4	6.7x
70	28.0	2.6x	27.1	7.5x
80	23.1	3.4x	21.1	9.1x

the regularizing effect of low sparsity. With SquareHead KD, we can obtain FP32 models with 70% and 75% sparsities which have essentially no loss (in terms of test accuracy) relative to the dense model, even though pruning is performed in one shot. We emphasize that fine-tuning with SquareHead KD improves the dense model's accuracy as well, from 28.2 to 33.0.

We now examine the speedup-vs-accuracy trade-offs, presented in Table 6.5 for both FP32 and INT8 models. In FP32, moderate sparsities (60–70%) can be reached losslessly, leading to speedups of 2–2.5x. Moving to INT8, we observe a consistent accuracy decrease of 1–2% at each sparsity level, due to post-training quantization. At the same time, this loss of accuracy is accompanied by a major performance improvement, since the gains due to these two compression approaches compound. In absolute terms, the 70% INT8 model can execute at a remarkable 7.7 tokens/

second on a single core of an AMD Ryzen CPU, and at 20.9 tokens/second on 4 cores. The speedup of the lossless 60%-sparse INT8 model is approximately 6.7x, and we can reach a decoding speedup of 9.08x at 80% INT8, at the price of a 7% drop in accuracy (Table 6.6).

Extension to Llama-2. We also investigate compression of the more recent and popular Llama-2 (Touvron et al. 2023) model of 7B size. We follow the same experimental setup as described for MPT-7B model: one-shot pruning with SparseGPT followed by sparse-finetuning with different variants of loss functions. As shown in Fig. 6.7, the proposed SquareHead knowledge distillation consistently outperforms the standard cross-entropy and KL-divergence loss functions.

Table 6.6 Performance (tokens/second) for MPT-7B at various sparsities, using DeepSparse on 1 and 4 cores, using Intel and AMD CPUs

Precision	Sparsity (%)	Xeon Gold 6430 # cores		Ryzen 9 7950X # cores	
		1	4	1	4
FP32	0	0.6	2.1	1.4	2.5
	70	1.5	5.5	2.0	6.3
	80	2.1	8.2	2.1	7.6
INT8	70	4.7	16.3	7.7	20.9
	80	6.1	19.6	7.9	26.7

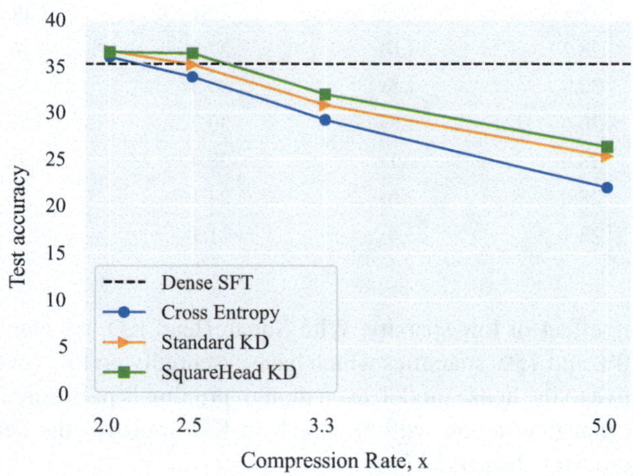

Fig. 6.7 Test accuracy for variants of loss functions on the GSM8K dataset and Llama-2-7B model at various compression rates

6.5 Discussion

We have shown results suggesting that sparsity can be an effective acceleration approach for LLM inference, focusing on applications such as speech and language translation, and text generation following complex instructions. Importantly, our study shows that sparsity is compatible with quantization in the memory-bound generative setting and that together these techniques can lead to remarkable performance for computationally-limited devices. Based on prior work, we have identified a general distillation approach to recover accuracy while inducing high sparsity over limited fine-tuning data. Future work could expand this study by exploring sparse fine-tuning for larger-scale models and tasks in the generative setting, higher quantization degrees, as well as the practically relevant setting of pruning on the pre-training task (Frantar et al. 2024), followed by fine-tuning the already-sparsified model on a specialized dataset.

6.6 Reproducibility

To promote reproducibility of our results, we provide the following resources:

- Code for sparse fine-tuning of T5 and Whisper models can be found here: https://github.com/IST-DASLab/TACO4NLP.
- Code for sparse fine-tuning of GPT-type models can be found here: https://github.com/IST-DASLab/SparseFinetuning.
- MPT models are released at: https://sparsezoo.neuralmagic.com/?datasets=gsm8k&ungrouped=true.
- CPU speedup for generative inference can be reproduced by following the instructions at https://github.com/neuralmagic/deepsparse/tree/main/research/mpt.

Acknowledgements We would like to thank Eugenia Iofinova for useful comments on an earlier version of this draft, and Artur Niederfahrenhorst for useful suggestions regarding fine-tuning on the GSM8k dataset.

References

Ardila R, Branson M, Davis K, Kohler M, Meyer J, Henretty M, Morais R, Saunders L, Tyers F, Weber G (2020) Common voice: a massively-multilingual speech corpus. In: Proceedings of the twelfth language resources and evaluation conference, pp 4218–4222. https://aclanthology.org/2020.lrec-1.520

Bojar O, Buck C, Federmann C, Haddow B, Koehn P, Leveling J, Monz C, Pecina P, Post M, Saint-Amand H, Soricut R, Specia L, Tamchyna A (2014) Findings of the 2014 workshop on statistical machine translation. In: Proceedings of the ninth workshop on statistical machine translation, pp 12–58. https://aclanthology.org/W14-3302

Chee J, Cai Y, Kuleshov V, Sa CD (2024) Quip: 2-bit quantization of large language models with guarantees. In: Advances in neural information processing systems, vol 36

Cobbe K, Kosaraju V, Bavarian M, Chen M, Jun H, Kaiser L, Plappert M, Tworek J, Hilton J, Nakano R, et al. (2021) Training verifiers to solve math word problems. arXiv preprint arXiv: 2110.14168

Dao T, Fu D, Ermon S, Rudra A, Ré C (2022) FlashAttention: fast and memory-efficient exact attention with IO-awareness. Adv Neural Inf Process Syst 35:16344–16359

Dettmers T, Zettlemoyer L (2023) The case for 4-bit precision: k-bit inference scaling laws. In: International conference on machine learning, pp 7750–7774

Dettmers T, Lewis M, Belkada Y, Zettlemoyer L (2022a) Llm.int8(): 8-bit matrix multiplication for transformers at scale. In: Advances in neural information processing systems 35: annual conference on neural information processing systems 2022, NeurIPS 2022

Dettmers T, Lewis M, Shleifer S, Zettlemoyer L (2022b) 8-bit optimizers via block-wise quantization. In: 9th International conference on learning representations, ICLR

Dettmers T, Svirschevski R, Egiazarian V, Kuznedelev D, Frantar E, Ashkboos S, Borzunov A, Hoefler T, Alistarh D (2024) Spqr: a sparse-quantized representation for near-lossless llm weight compression. In: The twelfth international conference on learning representations

Devlin J, Chang MW, Lee K, Toutanova K (2019) Bert: pre-training of deep bidirectional transformers for language understanding. In: North American chapter of the association for computational linguistics (NAACL)

Frantar E, Alistarh D (2022) Spdy: accurate pruning with speedup guarantees. In: International conference on machine learning, pp 6726–6743

Frantar E, Alistarh D (2023) Sparsegpt: massive language models can be accurately pruned in one-shot. In: International conference on machine learning, pp 10323–10337

Frantar E, Ashkboos S, Hoefler T, Alistarh D (2022) Gptq: accurate post-training quantization for generative pre-trained transformers. arXiv preprint arXiv:2210.17323

Frantar E, Ruiz C, Houlsby N, Alistarh D, Evci U (2024) Scaling laws for sparsely-connected foundation models. In: The twelfth international conference on learning representations

Gao L, Tow J, Abbasi B, Biderman S, Black S, DiPofi A, Foster C, Golding L, Hsu J, Le Noac'h A et al. (2023) A framework for few-shot language model evaluation. https://www.zenodoorg/records/102568367

Hinton G, Vinyals O, Dean J (2015) Distilling the knowledge in a neural network. arXiv preprint arXiv:1503.02531

Kurtic E, Campos D, Nguyen T, Frantar E, Kurtz M, Fineran B, Goin M, Alistarh D (2022) The optimal bert surgeon: scalable and accurate second-order pruning for large language models. In: Proceedings of the 2022 conference on empirical methods in natural language processing, pp 4163–4181

Kurtić E, Frantar E, Alistarh D (2024) Ziplm: inference-aware structured pruning of language models. In: Advances in neural information processing systems, vol 36

Kurtz M, Kopinsky J, Gelashvili R, Matveev A, Carr J, Goin M, Leiserson W, Moore S, Shavit N, Alistarh D (2020) Inducing and exploiting activation sparsity for fast inference on deep neural networks. In: International conference on machine learning, pp 5533–5543

LeCun Y, Denker J, Solla S (1989) Optimal brain damage. In: Advances in neural information processing systems, vol 2

Levenshtein V (1966) Binary codes capable of correcting deletions, insertions, and reversals. Soviet Phys Doklady 10:707–710

Li Q, Zhang Y, Li L, Yao P, Zhang B, Chu X, Sun Y, Du L, Xie Y (2023) Fptq: Fine-grained post-training quantization for large language models. arXiv preprint arXiv:2308.15987

Macháček M, Bojar O (2014) Results of the wmt14 metrics shared task. In: Proceedings of the ninth workshop on statistical machine translation, pp 293–301

NeuralMagic (2021) Deepsparse. https://github.com/neuralmagic/deepsparse

Papineni K, Roukos S, Ward T, Zhu WJ (2002) Bleu: a method for automatic evaluation of machine translation. In: Proceedings of the 40th annual meeting of the association for computational linguistics, pp 311–318

Radford A, Kim JW, Xu T, Brockman G, McLeavey C, Sutskever I (2023) Robust speech recognition via large-scale weak supervision. In: International conference on machine learning, pp 28492–28518

Raffel C, Shazeer N, Roberts A, Lee K, Narang S, Matena M, Zhou Y, Li W, Liu PJ (2020) Exploring the limits of transfer learning with a unified text-to-text transformer. J Mach Learn Res 5485–5551

Sanh V, Wolf T, Rush A (2020) Movement pruning: adaptive sparsity by fine-tuning. Adv Neural Inf Process Syst 33:20378–20389

Shao W, Chen M, Zhang Z, Xu P, Zhao L, Li Z, Zhang K, Gao P, Qiao Y, Luo P (2023) Omniquant: omnidirectionally calibrated quantization for large language models. In: The twelfth international conference on learning representations

Sun S, Cheng Y, Gan Z, Liu J (2019) Patient knowledge distillation for bert model compression. In: Proceedings of the 2019 conference on empirical methods in natural language processing and the 9th international joint conference on natural language processing (EMNLP-IJCNLP), pp 4323–4332

Team MN, et al. (2023) Introducing mpt-7b: a new standard for open-source, commercially usable llms

Touvron H, Martin L, Stone K, Albert P, Almahairi A, Babaei Y, Bashlykov N, Batra S, Bhargava P, Bhosale S et al. (2023) Llama 2: open foundation and fine-tuned chat models. arXiv preprint arXiv:2307.09288

Vaswani A, Shazeer NM, Parmar N, Uszkoreit J, Jones L, Gomez AN, Kaiser L, Polosukhin I (2017) Attention is all you need. In: Neural information processing systems. https://api.semanticscholar.org/CorpusID:13756489

Wei J, Bosma M, Zhao VY, Guu K, Yu AW, Lester B, Du N, Dai AM, Le QV (2021) Finetuned language models are zero-shot learners. arXiv preprint arXiv:2109.01652

Xia M, Zhong Z, Chen D (2022) Structured pruning learns compact and accurate models. In: Proceedings of the 60th annual meeting of the association for computational linguistics (Volume 1: Long papers). Association for Computational Linguistics, Dublin, Ireland, pp 1513–1528. https://aclanthology.org/2022.acl-long.107

Yao Z, Yazdani Aminabadi R, Zhang M, Wu X, Li C, He Y (2022) Zeroquant: Efficient and affordable post-training quantization for large-scale transformers. In: Advances in neural information processing systems, vol 35. Curran Associates, Inc., pp 27168–27183. https://proceedings.neurips.cc/paper_files/paper/2022/file/adf7fa39d65e2983d724ff7da57f00ac-Paper-Conference.pdf

Part IV
Sequence Efficiency and Model Compression

Chapter 7
TCNCA: Temporal CNN with Chunked Attention for Efficient Training on Long Sequences

Aleksandar Terzić, Michael Hersche, Geethan Karunaratne, Abu Sebastian, and Abbas Rahimi

Abstract MEGA is a recent Transformer-based neural network utilizing a linear recurrent architecture whose computation can be parallelized, a useful property during training as well as sequence encoding tasks. The parallel computation is based on the fast Fourier transform and scales as $O(LlogL)$, with L being the sequence length. We replace the linear recurrence in MEGA with a temporal convolutional network (TCN) which permits a large receptive field size with few TCN layers, and reduces the computational complexity to $O(L)$. We call the resulting model **TCNCA**, a Temporal Convolutional Network with Chunked Attention. We evaluate TCNCA on the tasks of EnWik8 language modelling, associative recall, a synthetic reasoning benchmark, and long-range-arena (LRA) sequence classification, and observe consistent improvements compared to MEGA in terms of both task-specific metrics and runtimes. More specifically, on EnWik8, TCNCA outperforms MEGA-chunk by a 0.01 BPC loss with a 1.22×/1.28× faster forward/backward pass. On LRA, TCNCA outperforms MEGA-chunk by 0.8% on average with a 1.42×/1.16× forward/backward pass speed-up. We further demonstrate the efficacy of our approach by comparing the runtimes of our approach and MEGA over a wide range of sequence lengths and embedding dimensions.

7.1 Introduction

Transformer (Vaswani et al. 2017) is a neural architecture which has found success in a variety of tasks including image processing (Dosovitskiy et al. 2021; Liu et al. 2021), physical system modelling (Geneva and Zabaras 2022), but perhaps most

A. Terzić
ETH Zurich, Zürich, Switzerland

A. Terzić (✉) · M. Hersche · G. Karunaratne · A. Sebastian · A. Rahimi
IBM Research, Zurich, Rüschlikon, Switzerland
e-mail: aleksandar.terzic1@ibm.com

© The Author(s), under exclusive license to Springer Nature Switzerland AG 2025
P. Passban et al. (eds.), *Enhancing LLM Performance*, Machine Translation: Technologies and Applications 7, https://doi.org/10.1007/978-3-031-85747-8_7

notably, language modelling (Brown et al. 2020; Chowdhery et al. 2023; Touvron et al. 2023a, b). However, despite its wide adaptation, it faces some crucial limitations.

The core limitation of the Transformer is its $O(L^2)$ computational and memory complexity, where L is the sequence length. The quadratic complexity arises as a consequence of the fact that computing *attention*, the Transformer's backbone, involves calculating the dot product between linear projections of all pairs of elements in the input sequence. This unfavourable property has spurred a large number of models proposing efficient approximations to the original Transformer (see Tay et al. (2020) for more details).

Another limitation of the Transformer is its poor performance on long-context classification tasks. This limitation of the Transformer has become clear with the introduction of a new family of artificial neural networks based on linear recurrences (Gu et al. 2022b; Orvieto et al. 2023; Smith et al. 2023). These recurrence-based alternatives outperform Transformer-based models by a large margin (Gu et al. 2022b) on the long-range-arena (LRA) benchmark (Tay et al. 2021)—a set of long-sequence classification tasks originally designed to test the performance of attention-based models. Attention is, however, still an important operation in LLMs, as there exists evidence that it is necessary in order to achieve state-of-the-art performance (Dao et al. 2023; Vardasbi et al. 2023). Note, however, that Gu and Dao (2023) question this claim.

Motivated by the current limitations, MEGA (Ma et al. 2023) combines the strengths of linear recurrences and attention in a manner which scales sub-quadratically in the sequence length. Concretely, MEGA combines a specific form of a linear recurrence, damped exponential moving average (EMA) (McKenzie and Gardner 2010), with chunked attention, which operates on fixed-size non-overlapping blocks in the input sequence. An overview of the model is given in Fig. 7.1 and more details to the EMA layer can be found in Sect. 7.2.

MEGA achieves competitive scores in a range of disparate tasks including language modelling on EnWik8 (Hutter 2006 and LRA sequence classification (Tay et al. 2021). MEGA is particularly interesting for us as it is the only model which has simultaneously demonstrated strong performance on both of these types of tasks.

Although MEGA is a powerful and efficient model, we can further improve both its modelling power and computational efficiency by considering alternative operations to the linear recurrence MEGA employs. In this research work we investigate the performance and runtime effects of replacing the bottleneck linear recurrence within the MEGA processing stack with a temporal convolutional neural network (TCN) (Bai et al. 2018; Kalchbrenner et al. 2016; van den Oord et al. 2016; Ingolfsson et al. 2020), an operator which scales linearly with the sequence length. The TCN employs dilated convolutions which allow the network to achieve a large receptive field with fewer parameters than a dense convolution would require. TCNs are typically implemented as a cascade of *residual blocks* (Ingolfsson et al. 2020), in which each block applies a dilated convolution on the input sequence, with the dilation exponentially increasing with each successive block (Bai et al. 2018; Ingolfsson et al. 2020). We call our resulting model, which combines a TCN with chunked attention, TCNCA.

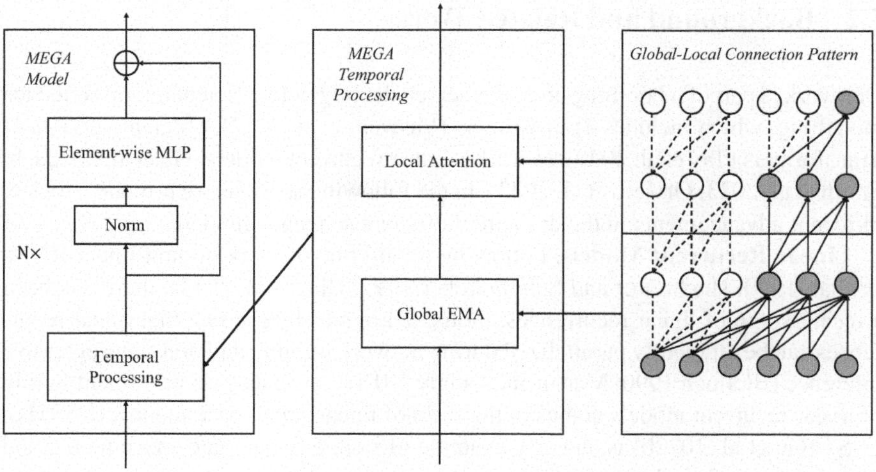

Fig. 7.1 A high-level overview of MEGA (Ma et al. 2023). MEGA consists of N layers, each of which contains a temporal processing block and a position-wise multi-layer perceptron (MLP). The temporal processing block in turn contains an EMA layer which operates over the entire input sequence and an attention layer which operates on fixed-size non-overlapping segments of the input sequence. Due to the global-local structure of temporal processing, each output token is influenced by each input token, a concept which we illustrate on the right-hand side of the figure. As shown, the final output token is impacted by all of the input tokens through the intermediate values which connect the global EMA with local attention. For a detailed overview of the temporal processing block, see Fig. 7.3

We evaluate TCNCA on EnWik8 language modelling, a synthetic benchmark, associative recall (Ba et al. 2016; Dao et al. 2023), as well as LRA long sequence classification. We find that on EnWik8 language modelling, TCNCA outperforms MEGA (and Transformer-XL Dai et al. 2019), achieving a bit-per-character (BPC) loss of 1.01, in addition to a 1.22×/1.28× faster forward/backward pass compared to MEGA-chunk. On a synthetic reasoning benchmark, *associative recall*, TCNCA is competitive with MEGA over a range of different sequence lengths and vocabulary sizes. On the LRA classification tasks, TCNCA outperforms MEGA-chunk by an average of 0.8% while achieving an average of a 1.42×/1.16× forward/backward pass speed-up. We further experiment with simpler versions of TCNCA on a subset of LRA tasks as well as on EnWik8.

Overall, we find TCNCA to be an efficient alternative to MEGA, offering benefits in terms of performance as well as increased efficacy, in particular during training on long sequences.

7.2 Background and Related Work

Our work is part of a booming area of research dealing with efficient neural sequence modelling, which includes Transformers (Vaswani et al. 2017), efficient variants of Transformers (Tay et al. 2020), as well as linear recurrent models (Gu et al. 2022a, b; Smith et al. 2023; Orvieto et al. 2023). In the following sections we provide a review of recent advancements in the domain of efficient sequence models.

Linear Recurrent Models. Following a long line of work on non-linear RNNs (Elman 1990; Hochreiter and Schmidhuber 1997; Cho et al. 2014), there has been a recent surge of linear recurrences, motivated in part by the fact that linear recurrences can be efficiently parallelized during network training and non-autoregressive inference (Blelloch 1990; Martin and Cundy 2018). Particularly, a prominent family of linear recurrent models comes in the form of linear state-space models (LSSMs).

S4 (Gu et al. 2022b) is an early example of such a linear state-space model and is, to the best of our knowledge, the first method to significantly advance the state-of-the-art on LRA classification (Tay et al. 2021). Many other variants of linear state-space models follow, including S4D (Gu et al. 2022a) that diagonalizes the recurrence matrix from S4, GSS (Mehta et al. 2022) that introduces a gating mechanism for improving the performance of LSSMs on language modelling, and S5 (Smith et al. 2023) that introduces multiple-input-multiple-output LSSMs. The LRU model (Orvieto et al. 2023) is a linear recurrent model which, in contrast to the LSSM family of models, is not based on discretizing an implicit continuous state-space model, while still achieving near-state-of-the-art accuracy on LRA.

Linear recurrent models map a 1-dimensional sequence u_t with $t = 1, ..., L$ to a 1-dimensional sequence y_t through a hidden h-dimensional vector x_t using the set of linear equations in (7.1):

$$\begin{aligned} x_t &= Ax_{t-1} + Bu_t \\ y_t &= Cx_t + Du_t \end{aligned} \tag{7.1}$$

with $A \in \mathbb{R}^{h\times h}$, $B, C^T \in \mathbb{R}^{h\times 1}$ and $D \in \mathbb{R}$. The set of equations is typically applied on each embedding dimensions separately, with a separate set of parameters A, B, C and D.

Equation (7.1) defines a linear time-invariant system and as such admits a different interpretation, namely, it is equivalent to a convolution of the input sequence x_t with a kernel of length L whose elements can be computed as $K_t = CA^tB$. The kernel can be efficiently computed when the A-matrix is diagonal, which is indeed the recent trend in linear recurrent models (Gu et al. 2022a; Ma et al. 2023). The convolution with the long kernel can be computed directly with a cost of $O(dL^2)$, but is computed more efficiently for long sequences using the fast Fourier transform (FFT) algorithm, whose computational cost is $O(dLlogL)$. We will assume that $L >> d$, making this cost asymptotically equal to $O(LlogL)$. An overview of the kernel generation and FFT-convolution can be found in Gu et al. (2022b) and Ma et al. (2023).

MEGA (Ma et al. 2023), the model which forms the basis of our work, utilizes a specific form of the linear equation (7.1) called "damped EMA", as described in Eq. (7.2):

$$\begin{aligned} x_t &= (1 - \alpha \odot \delta) \odot x_{t-1} + \alpha \odot Bu_t \\ y_t &= Cx_t + Du_t \end{aligned} \tag{7.2}$$

with $w_{\alpha,\delta} \in \mathbb{R}^h$ being trainable parameters, $\sigma(\cdot)$ denoting the sigmoid activation function, and $\alpha = \sigma(w_\alpha)$ $\delta = \sigma(w_\delta)$. Such a recurrence corresponds to a convolution kernel which exhibits an exponentially decaying shape.

Long Convolution Models. We refer to models which apply convolutions whose kernels are of the same length as the input sequence as *long convolutional models*. These models differ from linear recurrent models in several ways. Most notably, they do not necessarily admit a recurrent formulation. SGConv (Li et al. 2023), a member of this group of models, constructs a sparsely parametrized long kernel with an exponentially decaying structure and finds that this achieves strong performance on LRA. Hyena hierarchy (Poli et al. 2023) achieves strong scores on language modelling with a long convolutional kernel generated by a hypernetwork. FlashButterfly (Fu et al. 2023) explores simple long convolution kernel constructions that are effective on LRA classification and furthermore develops a hardware-aware algorithm for efficient computation of FFT-based long convolutions.

TCNs for Sequence Modelling. Temporal convolutional networks (TCNs) are widely used for parameter-efficient processing of data with long-range dependencies. CDIL-CNN (Cheng et al. 2022) employs circular dilated convolutions on several datasets including LRA. WaveNet (van den Oord et al. 2016) is a generative audio model based on the TCN architecture. ByteNet (Kalchbrenner et al. 2016) employs the TCN for machine translation. TCAN (Hao et al. 2020) employs a cascade of dilated convolutions with full self-attention. Their architecture differs from TCNCA in several ways, and scales quadratically with the sequence length. TConvTransformer (Chao et al. 2023) is a quadratic complexity concatenation of a TCN and multi-head self-attention.

7.3 The TCNCA Model

A high-level overview of TCNCA is shown in Fig. 7.2 and a more detailed view of the model construction is provided in Fig. 7.3.

TCNCA consists of N layers, each of which alternates a temporal processing block with an element-wise MLP. The layers are additionally augmented with normalization and residual connections, closely following the original Transformer architecture. Each temporal processing block combines a TCN with attention, where the TCN operates over the entire input sequence (global), and attention is restricted to fixed-size chunks of the sequence (local).

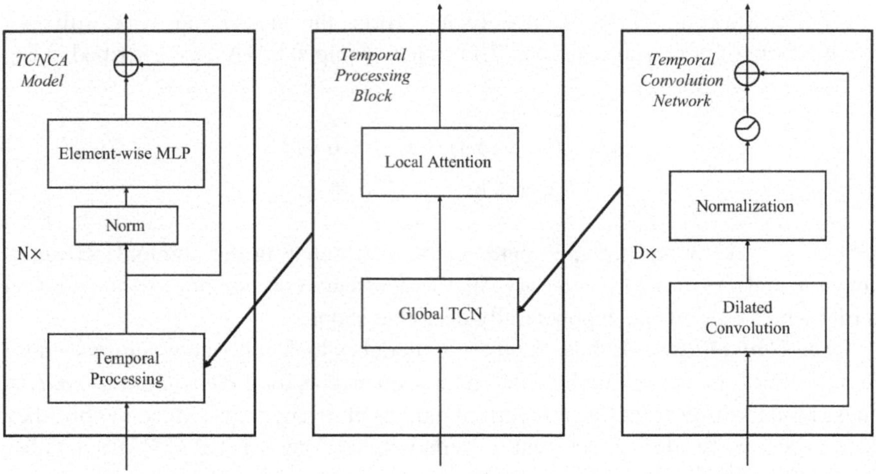

Fig. 7.2 A high-level overview of TCNCA. TCNCA relies heavily on the MEGA architecture. It replaces the EMA module with a temporal convolutional network (TCN), which is composed of a cascade of residual blocks of depth D, each of which includes a dilated convolution operation. Within every single TCN block, the dilation increases exponentially with the depth of the corresponding residual block. The temporal processing block visualized here is a simplified view of the full block, shown in Fig. 7.3

The TCN is depicted on the right side of Fig. 7.2. It is constructed as a stack of residual blocks of depth D, each block in turn consisting of a dilated convolution, normalization, and a nonlinear activation function. In our architecture, the non-linearity is the SiLU activation function (Elfwing et al. 2018).

A separate set of convolutional filters is instantiated for each embedding dimension of the input, in the same manner as how linear recurrent networks operate with a separate set of parameters per embedding dimension (Gu et al. 2022a, b; Gu and Dao 2023). The amount of kernel dilation increases exponentially with the depth of the residual block, ranging from 0 to F^{D-1}, where F is an integer hyperparameter called the dilation factor. The computational cost of the TCN is $O(dL)$, with d being the embedding dimension. This is $O(L)$ when $L >> d$. When the task is sequence encoding, meaning a non-autoregressive task such as classification, the input sequence is circularly padded (as in Cheng et al. (2022)). During decoding, meaning autoregressive tasks such as language modelling, the input sequence is zero-padded on the left.

Following the global TCN, local (chunked) attention operates on fixed size blocks of the input sequence. Concretely, given an input sequence $X_{1,\ldots,L}$, chunked attention splits the sequence into non-overlapping chunks of size $C < L$ and computes attention within each chunk separately. The attention module is not the usual multi-head self-attention, but is instead a single-head gated attention module inspired by the gated attention unit (Hua et al. 2022). The computational cost of chunked attention is $O(dLC)$, which becomes $O(L)$ for very long sequences.

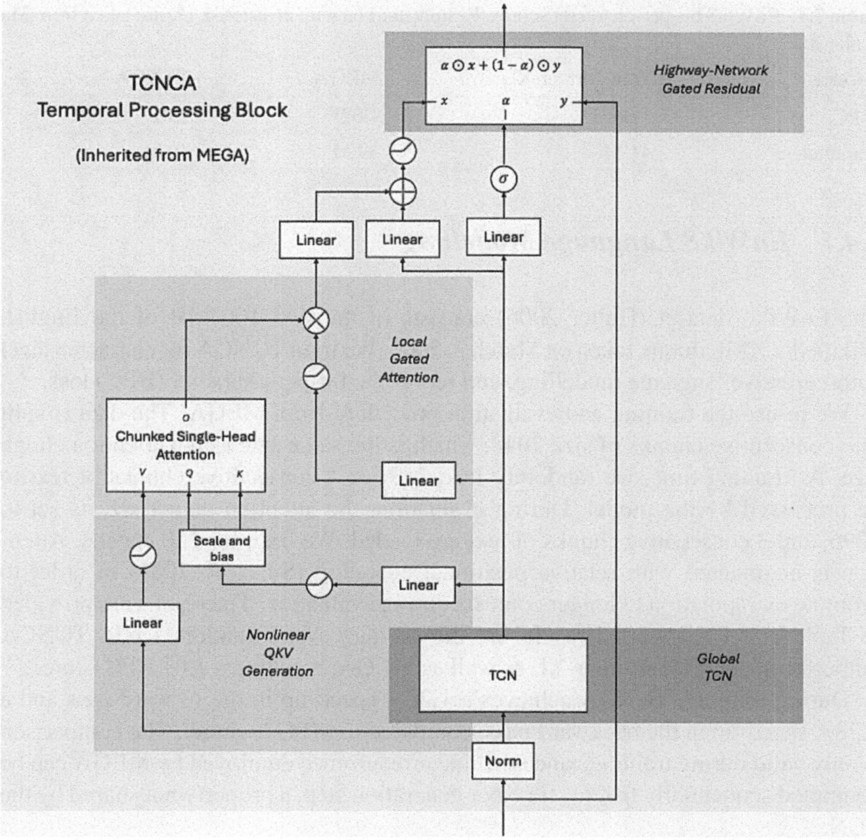

Fig. 7.3 The temporal processing block from Fig. 7.2 is in fact defined according to the architecture shown above, taken as is from MEGA (Ma et al. 2023) with the exception that MEGA's EMA layer is replaced by a TCN. It involves several steps which deviate from the usual construction of Transformer-based networks. The main differences from the usual construction are highlighted in the figure, and include: a global temporal processing layer, non-linear query, key and value generation, gated single-head chunked attention, and finally a type of gated residual connection originally introduced as a Highway network (Srivastava et al. 2015). The main results were obtained using this architecture. However, for comparison purposes results from other simpler alternatives are provided too

7.4 Experiments

Our evaluations are centered on language modelling with real and synthetic data and long sequence classification. In our evaluation, we observe a consistent improvement in task-specific metrics as well as processing run-times when comparing TCNCA to the chunked version of MEGA (Ma et al. 2023).

Table 7.1 EnWik8 bit-per-character scores. Results marked with an asterisk (*) are taken from Ma et al. (2023)

Model	Transformer-XL	MEGA	TCNCA
BPC	1.06*	1.02*	**1.01**
Parameters	41 M	39 M	39 M

7.4.1 EnWik8 Language Modelling

The EnWik8 dataset (Hutter 2006) consists of the first 100 MB of the English Wikipedia XML dump, taken on March 3, 2006. We train TCNCA for character-level autoregressive language modelling, and report the bit-per-character (BPC) loss.

We re-use the training and evaluation procedure from MEGA. The data is split into consecutive chunks of size 2048, which is the same size as the attention chunk size. At training time, we randomly load 2, 3, or 4 consecutive chunks of text to be processed by the model. During evaluation, the attention chunk size is set to 4096, and 3 consecutive chunks of text are loaded. We train for 250 epochs. Attention is augmented with relative positional encoding (Su et al. 2021) in order to promote extrapolation to longer contexts during evaluation. The results are provided in Table 7.1. The reported results are the average over 3 random seeds. TCNCA outperforms the Transformer-XL as well as MEGA, reaching a 1.01 BPC score.

During training, TCNCA achieves a $1.22\times$ speed-up in the forward pass and a $1.28\times$ speed-up in the backward pass, compared to MEGA-chunk. The comparison is only valid during training, since the linear recurrence employed by MEGA can be computed sequentially in $O(1)$ time per generation step, a property not shared by the TCN.

7.4.2 Associative Recall

Associative recall (Ba et al. 2016; Dao et al. 2023) is a synthetic reasoning benchmark which measures a basic reasoning capability of neural sequence models, remembering associations between pairs of tokens. There exists evidence that a model's performance on associative recall and more complex variants thereof is correlated with its performance on more general language modelling tasks (Arora et al. 2023; Dao et al. 2023; Olsson et al. 2022).

In this task, the input sequence consists of pairs of characters, the first one in the pair denoting a *key* and the second one being the associated *value*. After observing a sequence of key-value pairs, the model is prompted with a key and is required to output the corresponding value. An example would be:

Input: a1b2c3d4b

Output: 2

Table 7.2 Associative recall accuracy (%) with varying sequence lengths and vocabulary sizes

	Vocabulary size 10		Vocabulary size 20	
Sequence length	MEGA	TCNCA	MEGA	TCNCA
64	100.0	100.0	64.4	63.6
1024	100.0	100.0	99.4	99.8
4096	100.0	100.0	100.0	100.0

We test our model on examples with three different sequence lengths and two different vocabulary sizes. For sequence length 64, we use an attention chunk size of 32. For all other sequence lengths, a chunk size of 128 is used. The embedding dimension is 128 for all configurations. The best results over 10 random seeds are reported in Table 7.2. We see that over the investigated range of vocabulary sizes and sequence lengths, TCNCA remains competitive with MEGA.

7.4.3 Long-Range-Arena

Long-range-arena (Tay et al. 2021) comprises six classification tasks with sequence lengths ranging from 1024 to 16384. The benchmarks are varied, including pattern detection, sentiment classification, mathematical reasoning, and visual reasoning. Results on this task, averaged over 3 random seeds, are shown in Table 7.3.

TCNCA achieves a modest 0.8% increase in average accuracy compared to MEGA while offering a forward/backward pass speed-up in all benchmarks excluding ListOps, with average speed-ups being 42% in the forward pass and 16% in the backward pass. Furthermore, TCNCA outperforms every other model used in

Table 7.3 Long-range-arena accuracies (%) of state-of-the-art models. The Transformer scores are taken from the reproduction in MEGA. All other results, excluding TCNCA, were taken from the respective papers. The last row reports the end-to-end inference speed-up of TCNCA measured against MEGAchunk

Model	ListOps	Text	Retrieval	Image	Path	Path-X	Average
Transformer (Ma et al. 2023)	37.1	65.2	79.1	42.9	71.8	50	57.7
S4D (Gu et al. 2022a)	60.5	86.2	89.5	89.9	93.1	91.9	85.2
S5 (Smith et al. 2023)	62.2	89.3	91.4	90.1	95.3	98.6	87.8
LRU (Orvieto et al. 2023)	60.2	89.4	89.9	89.0	95.7	96.0	86.7
SGConv (Li et al. 2023)	61.4	89.2	91.1	88.0	95.4	97.8	87.1
MEGA chunk (Ma et al. 2023)	58.7	90.2	91.0	85.8	94.4	93.8	85.6
TCNCA	59.4	90.0	89.5	90.4	94.7	94.4	86.4
Speedup (forward)	1.12×	1.28×	1.73×	1.42×	1.30×	1.68×	1.42×
Speedup (backward)	0.95×	1.05×	1.34×	1.18×	1.03×	1.41×	1.16×

our comparison on the sequential CIFAR-10 image classification task (Krizhevsky and Hinton 2009), in which images are laid out in a 1-dimensional row instead of the usual 2-D grid.

7.4.4 Simpler TCNCA Variants

MEGA defines a complex temporal processing block which TCNCA inherits (Fig. 7.3). In this section, we investigate whether simpler TCNCA variants remain competitive with some of the baselines we have established in the previous sections. To this end, we define three simple TCNCA variants shown in Fig. 7.4, and evaluate them on a subset of LRA and on EnWik8.

The TCN hyperparameters (kernel size, dilation factor, TCN depth) are the same as those used to obtain the results shown in the previous sections. For each benchmark, we train with three different learning rates, $0.1\times$, $1\times$ and $5\times$ the learning rate used in

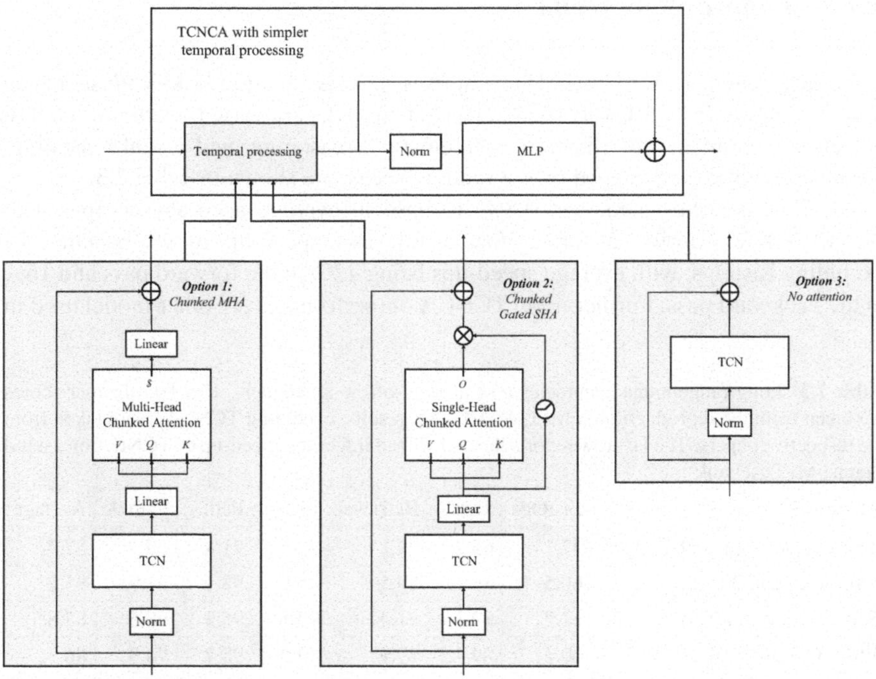

Fig. 7.4 We experiment on three simpler versions of TCNCA on a subset of LRA and EnWik8 language modelling. Option 1 (TCNCMHA: TCN with Chunked Multi-Head-Attention) implements temporal processing as a global TCN whose output is fed into multi-head chunked attention. Option 2 (TCNCSHGA: TCN with Chunked Single-Head Gated Attention) replaces multi-head attention from Option 1 with gated single-head attention, motivated by MEGA and the gated attention unit (Hua et al. 2022). Option 3 (TCN) fully removes attention

Table 7.4 Accuracy on a subset of LRA (higher is better), and BPC loss on EnWik8 language modelling (lower is better), for three different simple variants of TCNCA shown in Fig. 7.4

Model	ListOps ↑	Text ↑	Image ↑	Pathfinder ↑	Average ↑	EnWik8 ↓
TCNCA	59.4	90.0	90.4	94.4	83.6	1.01
TCNCMHA (1)	44.9	86.0	85.2	82.3	74.6	1.05
TCNCGSHA (2)	52.9	85.9	88.8	94.4	80.5	1.06
TCN (3)	52.2	89.1	88.9	96.3	81.6	1.27

the corresponding experiments (for experimental hyperparameters see Sect. 7.4.6). In the models which use attention, query, key and value dimensions are equal to the embedding dimension. Option 1, which utilizes multi-head attention, operates with 4 attention heads. In each case, the embedding dimension is selected such that the models are parameter-matched to TCNCA. The results are shown in Table 7.4. The reported results are the average over 3 random seeds.

On all benchmarks except Pathfinder, TCNCA outperforms all the simpler variants, confirming that the architectural choices defined in MEGA and inherited by our model are crucial. On the subset of LRA we experiment on, out of all of the simpler models, the one without attention (option 3) performed best. It outperforms TCNCA on Pathfinder by 1.9%, but its average performance on LRA is lower by 2% than that of TCNCA. On EnWik8, however, the model without attention is not able to match the quality of all of the others which do contain attention, confirming that it is an important component for the applicability of TCNCA on language modelling tasks.

7.4.5 Runtime Measurements

The runtime measurements were collected using the PyTorch benchmark utilities[1] with PyTorch version 2.2.1, CUDA version 12.1. All measurements were collected on a single 32GB Nvidia V100 GPU and two Intel Xeon 6258R CPU cores running at 2.70GHz.

Figures 7.5, 7.6 and 7.7 show the most important trends in terms of model efficacy during autoregressive training and non-autoregressive training and inference. OOM entries stand for "out of memory". We measure the speed-ups of TCNCA compared to MEGA in both the forward and backward pass using different embedding dimensions and EMA hidden state sizes. We additionally measure the speed-ups in both autoregressive and non-autoregressive processing modes (when EMA is run in a non-autoregressive mode, its hidden state dimension is expanded by 2, meaning that different sets of parameters are used when processing the sequence in the two directions Ma et al. 2023). The TCN is always configured with 3 residual blocks as is typically used in our experiments, with the kernel size set such that MEGA and TCNCA are roughly parameter-matched. The receptive field size of the TCN

[1] https://pytorch.org/tutorials/recipes/recipes/benchmark.html.

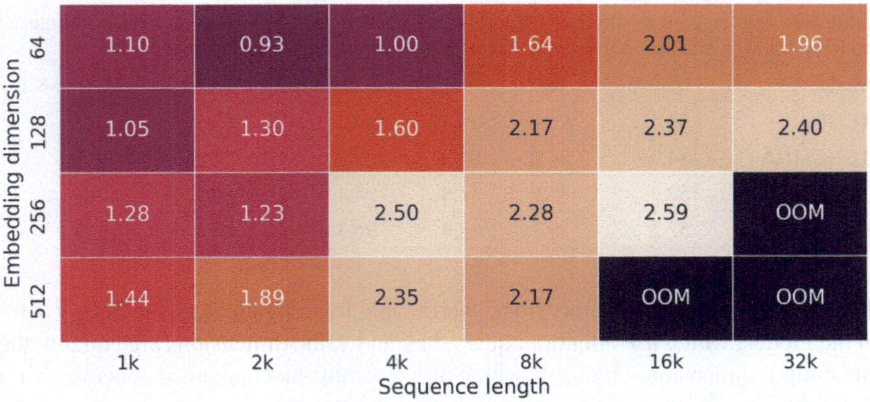

(a) Forward pass.

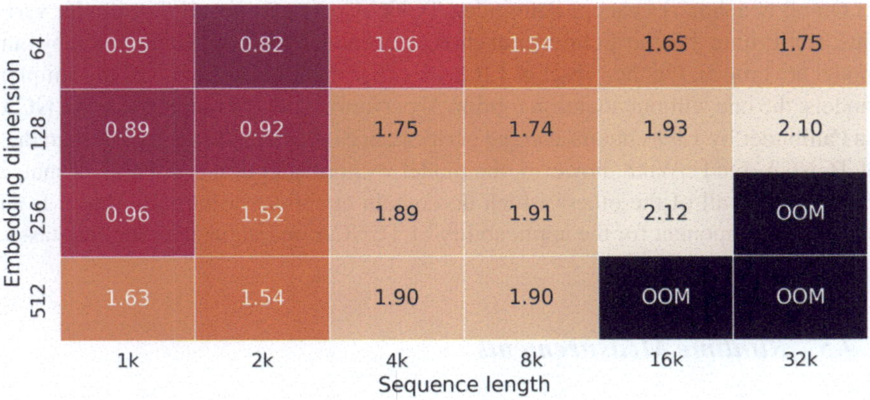

(b) Backward pass.

Fig. 7.5 Speedups of TCNCA over MEGA-chunk with an 8-layer model, EMA hidden state dimension 16, chunk size 1024, encoding (non-autoregressive) inference mode

is modified by changing the dilation factor, and is set such that the entire network covers the input sequence at least once.

The speedup increases with increased sequence length, EMA hidden state size, as well embedding dimension, indicating that TCNCA becomes a more attractive alternative to MEGA with larger model sizes and sequence lengths.

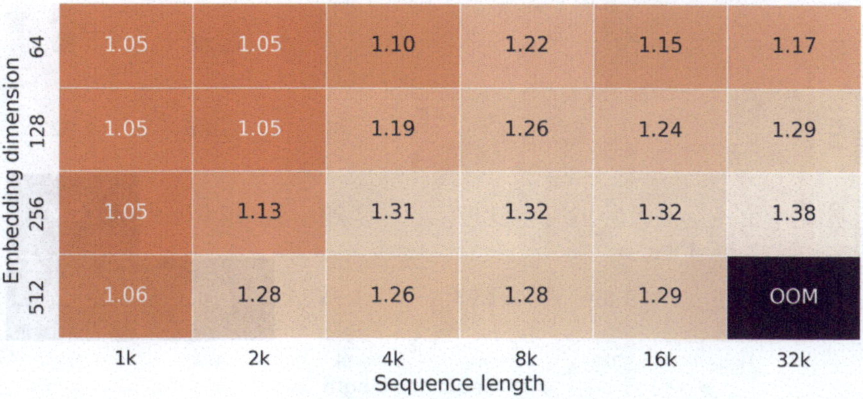

(a) Forward pass.

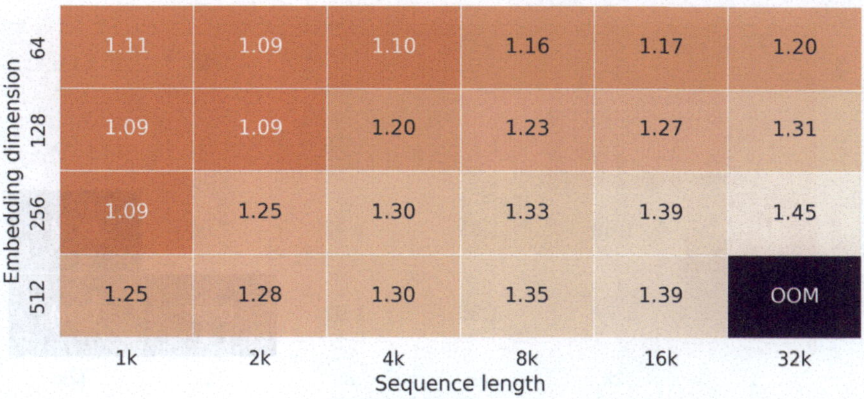

(b) Backward pass.

Fig. 7.6 Speedups of TCNCA over MEGA-chunk with an 8-layer model, EMA hidden state dimension 16, chunk size 1024, decoding (autoregressive) inference mode

7.4.6 Hyperparameters

The hyperparameters for reproducing our experimental results are provided in Tables 7.5, 7.6 and 7.7. K denotes the kernel size, F the dilation factor, and D the number of TCN residual blocks. In the associative recall experiments, the embedding dimension is 128, attention chunk size is 128, and the model consists of two decoder layers.

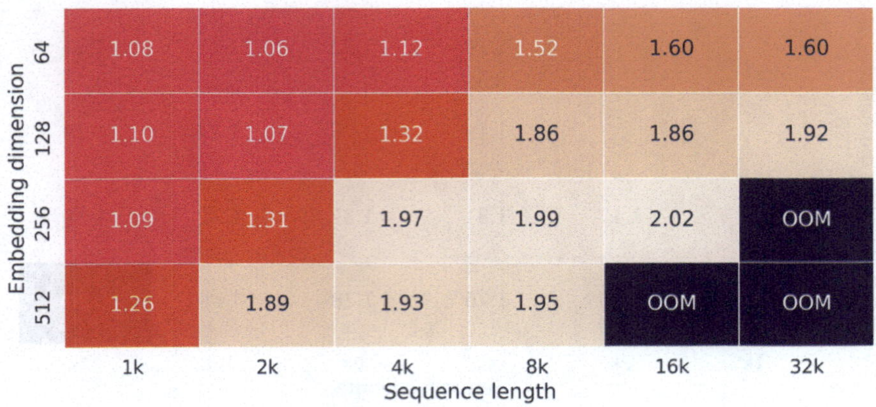

(a) Forward pass.

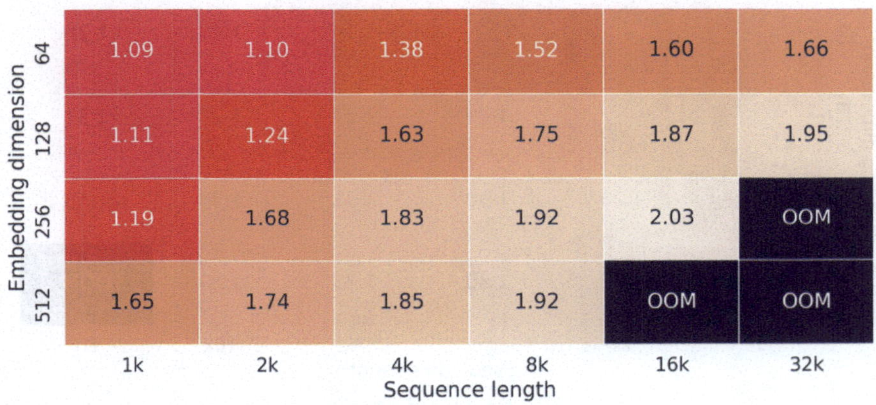

(b) Backward pass.

Fig. 7.7 Speedups of TCNCA over MEGA-chunk with an 8-layer model, EMA hidden state dimension 60, chunk size 1024, decoding (autoregressive) inference mode

Table 7.5 TCNCA hyperparameters for LRA and EnWik8. All other hyperparameters and architectural options are taken from MEGA

	K	F	D	Learning rate
ListOps	15	14	2	0.001
Text	7	4	3	0.004
Retrieval	11	8	3	0.0005
Image	11	8	3	0.01
Path	11	8	3	0.01
Path-X	11	8	3	0.001
EnWik8	15	12	3	0.005

Table 7.6 TCNCA hyperparameters for the associative recall experiments

Vocab size	Seq. length	LR	Dropout	Epochs	Batch size	K	F	D
10	64	0.001	0.1	800	128	15	2	3
	1024	0.01	0.1	800	128	15	8	3
	4096	0.01	0	800	128	15	14	3
20	64	0.01	0.1	800	128	15	2	3
	1024	0.01	0.1	800	128	15	8	3
	4096	0.012	0	2000	32	15	14	3

Table 7.7 MEGA hyperparameters for the associative recall experiments

Vocab size	Seq. length	LR	Dropout	Epochs	Batch size
10	64	0.01	0.1	800	128
	1024	0.01	0	800	128
	4096	0.01	0	800	128
20	64	0.01	0	800	128
	1024	0.008	0.1	2000	64
	4096	0.01	0.1	2000	64

7.5 Conclusion

In this work inspired by MEGA (Ma et al. 2023), we showed that a TCN and chunked attention hybrid model, TCNCA, is able to compete with the state-of-the-art models on Enwik8 language modelling and Long-Range-Arena sequence classification, while offering better performance and run-times compared to MEGA-chunk. Concretely, our model outperforms the competing method, MEGA-chunk, by 0.01 BPC points on EnWik8 language modelling while offering an over 20% speed-up in the forward and backward pass. We measure the performance of TCNCA on a synthetic language modelling benchmark, associative recall, and confirm that it performs on par with MEGA. On LRA classification, TCNCA outperforms MEGA-chunk by 0.8% while offering an average speedup of 46% in the forward pass and 16% in the backward pass. TCNCA achieves the best result of all competing methods on the Image task from LRA. We furthermore evaluated three simple variants of TCNCA and found that the original model formulation is the most performant one, which confirms that the architectural choices introduced by MEGA and re-used by TCNCA are well motivated. Our runtime measurements show that TCNCA is an efficient alternative to MEGA-chunk, in particular during training of large models on long sequences.

Our work is limited in several ways. Firstly, on language modelling, our model offers run-time advantages over MEGA during training but not during inference. Secondly, the linear recurrence in MEGA can be computed using an alternative algorithm to FFT, the *parallel scans* algorithm (Blelloch 1990), and it is unclear whether our model is still computationally advantageous in this scenario. Finally, our analysis in Sect. 7.4.4 confirms that MEGA's complex temporal processing block is performant, but it does not explore the effect of each component of the block in detail. All of these limitations are possible avenues for further research.

References

Arora S, Eyuboglu S, Timalsina A, Johnson I, Poli M, Zou J, Rudra A, Ré C (2023) Zoology: measuring and improving recall in efficient language models. arXiv preprint arXiv:2312.04927

Ba J, Hinton GE, Mnih V, Leibo JZ, Ionescu C (2016) Using fast weights to attend to the recent past. Adv Neural Inf Process Syst (NeurIPS) 29

Bai S, Kolter JZ, Koltun V (2018) An empirical evaluation of generic convolutional and recurrent networks for sequence modeling. arXiv:abs/1803.01271. https://api.semanticscholar.org/CorpusID:4747877

Blelloch GE (1990) Prefix sums and their applications. Carnegie Mellon University Pittsburgh, School of Computer Science, PA, USA

Brown T, Mann B, Ryder N, Subbiah M, Kaplan JD, Dhariwal P, Neelakantan A, Shyam P, Sastry G, Askell A, Agarwal S, Herbert-Voss A, Krueger G, Henighan T, Child R, Ramesh A, Ziegler D, Wu J, Winter C, Hesse C, Chen M, Sigler E, Litwin M, Gray S, Chess B, Clark J, Berner C, McCandlish S, Radford A, Sutskever I, Amodei D (2020) Language models are few-shot learners. In: Advances in neural information processing systems, curran associates, Inc., vol 33, pp 1877–1901. https://proceedings.neurips.cc/paper_files/paper/2020/file/1457c0d6bfcb4967418bfb8ac142f64a-Paper.pdf

Chao X, Ge Z, Leung H (2023) Video2mesh: 3d human pose and shape recovery by a temporal convolutional transformer network. IET Comput Vis

Cheng L, Khalitov R, Yu T, Zhang J, Yang Z (2022) Classification of long sequential data using circular dilated convolutional neural networks. Neurocomputing

Cho K, van Merriënboer B, Bahdanau D, Bengio Y (2014) On the properties of neural machine translation: encoder–decoder approaches. In: Proceedings of SSST-8, eighth workshop on syntax, semantics and structure in statistical translation. Doha, Qatar, pp 103–111

Chowdhery A, Narang S, Devlin J, Bosma M, Mishra G, Roberts A, Barham P, Chung HW, Sutton C, Gehrmann S et al (2023) Palm: scaling language modeling with pathways. J Mach Learn Res 24:1–113

Dai Z, Yang Z, Yang Y, Carbonell J, Le Q, Salakhutdinov R (2019) Transformer-XL: attentive language models beyond a fixed-length context. In: Proceedings of the 57th annual meeting of the association for computational linguistics (ACL), pp 2978–2988

Dao T, Fu DY, Saab KK, Thomas AW, Rudra A, Ré C (2023) Hungry Hungry Hippos: towards language modeling with state space models. In: International conference on learning representations (ICLR)

Dosovitskiy A, Beyer L, Kolesnikov A, Weissenborn D, Zhai X, Unterthiner T, Dehghani M, Minderer M, Heigold G, Gelly S, Uszkoreit J, Houlsby N (2021) An image is worth 16x16 words: transformers for image recognition at scale. In: International conference on learning representations (ICLR)

Elfwing S, Uchibe E, Doya K (2018) Sigmoid-weighted linear units for neural network function approximation in reinforcement learning. Neural Netw: Offic J Int Neural Netw Soc 107:3–11

Elman JL (1990) Finding structure in time. Cogn Sci 14:179–211

Fu DY, Epstein EL, Nguyen E, Thomas AW, Zhang M, Dao T, Rudra A, Re C (2023) Simple hardware-efficient long convolutions for sequence modeling. In: ICLR 2023 workshop on mathematical and empirical understanding of foundation models

Geneva N, Zabaras N (2022) Transformers for modeling physical systems. Neural Netw 272–289

Gu A, Dao T (2023) Mamba: linear-time sequence modeling with selective state spaces. arXiv preprint arXiv:2312.00752

Gu A, Goel K, Gupta A, Ré C (2022) On the parameterization and initialization of diagonal state space models. Adv Neural Inf Process Syst (NeurIPS) 35:35971–35983

Gu A, Goel K, Ré C (2022b) Efficiently modeling long sequences with structured state spaces. In: International conference on learning representations (ICLR)

Hao H, Wang Y, Xia Y, Zhao J, Shen F (2020) Temporal convolutional attention-based network for sequence modeling. arXiv preprint arXiv:2002.12530

Hochreiter S, Schmidhuber J (1997) Long short-term memory. Neural Comput 9(8):1735–1780

Hua W, Dai Z, Liu H, Le Q (2022) Transformer quality in linear time. In: International conference on machine learning (ICML), PMLR, pp 9099–9117

Hutter M (2006) The human knowledge compression contest. http://prize.hutter1.net/

Ingolfsson TM, Hersche M, Wang X, Kobayashi N, Cavigelli L, Benini L (2020) Eeg-tcnet: An accurate temporal convolutional network for embedded motor-imagery brain–machine interfaces. In: IEEE international conference on systems, man, and cybernetics (SMC), pp 2958–2965

Kalchbrenner N, Espeholt L, Simonyan K, Oord Avd, Graves A, Kavukcuoglu K (2016) Neural machine translation in linear time. arXiv preprint arXiv:1610.10099

Krizhevsky A, Hinton G (2009) Learning multiple layers of features from tiny images. Tecnical report 0, University of Toronto, Toronto, Ontario. https://www.cs.toronto.edu/~kriz/learning-features-2009-TR.pdf

Li Y, Cai T, Zhang Y, Chen D, Dey D (2023) What makes convolutional models great on long sequence modeling? In: International conference on learning representations (ICLR), Kigali, Rwanda

Liu Z, Lin Y, Cao Y, Hu H et al (2021) Swin transformer: hierarchical vision transformer using shifted windows. In: Proceedings of the IEEE/CVF international conference on computer vision, pp 10012–10022

Ma X, Zhou C, Kong X, He J, Gui L, Neubig G, May J, Luke Z (2023) Mega: moving average equipped gated attention. In: International conference on learning representations (ICLR), Kigali, Rwanda

Martin E, Cundy C (2018) Parallelizing linear recurrent neural nets over sequence length. In: International conference on learning representations (ICLR)

McKenzie E, Gardner ES (2010) Damped trend exponential smoothing: a modelling viewpoint. Int J Forecast 26(4):661–665

Mehta H, Gupta A, Cutkosky A, Neyshabur B (2022) Long range language modeling via gated state spaces. arXiv preprint arXiv:2206.13947

Olsson C, Elhage N, Nanda N, Joseph N, DasSarma N, Henighan T, Mann B, Askell A, Bai Y, Chen A et al (2022) In-context learning and induction heads. arXiv preprint arXiv:2209.11895

Orvieto A, Smith SL, Gu A, Fernando A, Gulcehre C, Pascanu R, De S (2023) Resurrecting recurrent neural networks for long sequences. In: International conference on machine learning (ICML)

Poli M, Massaroli S, Nguyen EQ, Fu DY, Dao T, Baccus SA, Bengio Y, Ermon S, Ré C (2023) Hyena hierarchy: towards larger convolutional language models. In: International conference on machine learning (ICML), Hawaii, US

Smith JT, Warrington A, Linderman S (2023) Simplified state space layers for sequence modeling. In: International conference on learning representations (ICLR), Kigali, Rwanda

Srivastava RK, Greff K, Schmidhuber J (2015) Training very deep networks. Adv Neural Inf Process Syst (NeurIPS) 28

Su J, Lu Y, Pan S, Murtadha A, Wen B, Liu Y (2021) Roformer: enhanced transformer with rotary position embedding. arXiv preprint arXiv:2104.09864

Tay Y, Dehghani M, Bahri D, Metzler D (2020) Efficient transformers: a survey. ACM Comput Surv 55:1–28

Tay Y, Dehghani M, Abnar S, Shen Y, Bahri D, Pham P, Rao J, Yang L, Ruder S, Metzler D (2021) Long range arena: a benchmark for efficient transformers. In: International conference on learning representations (ICLR)

Touvron H, Lavril T, Izacard G, Martinet X, Lachaux MA, Lacroix T, Rozière B, Goyal N, Hambro E, Azhar F et al (2023a) Llama: open and efficient foundation language models. arXiv preprint arXiv:2302.13971

Touvron H, Martin L, Stone K, Albert P, Almahairi A, Babaei Y, Bashlykov N, Batra S, Bhargava P, Bhosale S et al (2023b) Llama 2: open foundation and fine-tuned chat models. arXiv preprint arXiv:2307.09288

van den Oord A, Dieleman S, Zen H, Simonyan K, Vinyals O, Graves A, Kalchbrenner N, Senior A, Kavukcuoglu K (2016) WaveNet: a generative model for raw audio. In: Proceedings of the 9th ISCA workshop on speech synthesis workshop (SSW 9), p 125

Vardasbi A, Pires T, Schmidt RM, Peitz S (2023) State spaces aren't enough: machine translation needs attention. In: European association for machine translation conferences/workshops. https://api.semanticscholar.org/CorpusID:258309400

Vaswani A, Shazeer NM, Parmar N, Uszkoreit J, Jones L, Gomez AN, Kaiser L, Polosukhin I (2017) Attention is all you need. In: Neural information processing systems. https://api.semanticscholar.org/CorpusID:13756489

Chapter 8
Class-Based Feature Knowledge Distillation

Khouloud Saadi, Jelena Mitrović, and Michael Granitzer

Abstract Knowledge Distillation (KD) is an effective technique for compressing large language models through the teacher-student framework without compromising accuracy. Once a large model is compressed to a lightweight one, it can be stored within limited disk space and trained at higher speed with fewer computational resources. Previous work on feature distillation mainly applied an exact matching between the hidden representations of the student and the teacher. However, since the student has a lower capacity compared to the teacher, it may struggle to mimic its exact hidden representations. This leads to a large discrepancy between their features as demonstrated in preceding research. Therefore, we propose intra-class similarity-guided feature distillation, a novel approach to make the task easier for the student. In this chapter, we map each sample representation by the student to its k-nearest neighbor samples representations by the teacher that are within the same class. This method is novel and can be combined with other distillation techniques. Empirical results show the effectiveness of our proposed approach by maintaining strong performance on benchmark datasets. Furthermore, we evaluate our proposed KD approach on the hate-speech detection task, where we aim to compress HateBERT into a smaller, more efficient model.

8.1 Introduction

Knowledge Distillation (KD) (Romero et al. 2014; Hinton et al. 2015) is known as an effective technique to compress LLMs (Sanh et al. 2019; Sun et al. 2019; Jiao et al. 2020). It is a framework to train a student network, the model with fewer parameters, to mimic the behaviour of a teacher network, the over-parameterized model, on a set of data points. The distilled lightweight student model is highly efficient in terms

K. Saadi (✉) · J. Mitrović · M. Granitzer
University of Passau, Passau, Germany
e-mail: khouloud.saadi@uni-passau.de

J. Mitrović
Institute for Artificial Intelligence Research and Development of Serbia, Novi Sad, Serbia

© The Author(s), under exclusive license to Springer Nature Switzerland AG 2025
P. Passban et al. (eds.), *Enhancing LLM Performance*, Machine Translation: Technologies and Applications 7, https://doi.org/10.1007/978-3-031-85747-8_8

of parameters, requiring minimal storage and computational resources. There are different approaches to KD, where the teacher is dynamic as in Zhou et al. (2021) and Ma et al. (2022), or static as in Sun et al. (2019) or Jiao et al. (2020). The knowledge embedded in various components of the teacher can be distilled to the student. Examples include the prediction layer (Hinton et al. 2015; Sanh et al. 2019), the attention matrices (Jiao et al. 2020; Passban et al. 2021; Wang et al. 2021), and the hidden states (Sun et al. 2019; Jiao et al. 2020; Wu et al. 2020; Passban et al. 2021; Saadi et al. 2023). In Kovaleva et al. (2019), it is shown that LLMs (such as BERT) suffer from over-parametrization in domain-specific tasks. Thus, task-specific distillation has been an active research topic. In this chapter, we mainly focus on task-specific feature distillation from a static teacher.

Existing methods in feature distillation tried to improve the loss function where MSE (Sun et al. 2019; Jiao et al. 2020), cosine distance (Sanh et al. 2019), and correlation function (Saadi et al. 2023) are used to match the hidden representations of the teacher and the student. However, previous work mostly applied a one-to-one mapping between the student hidden representations and the teacher hidden representations (Sun et al. 2019; Sanh et al. 2019) neglecting the capacity gap between them. In fact, each student sample representation is mapped to the identical teacher sample representation. Nevertheless, as detailed in Chen et al. (2022), in layer-wise distillation, the student may struggle to mimic the hidden representations of the teacher because of their large capacity difference. This invariably results in huge discrepancies between their feature representations. Furthermore, as shown in Liang et al. (2023), training a student to achieve discriminative feature extraction for the main classification task and exact feature matching for distillation at the same time is considered multi-task learning. It is also shown that, in this case, the student tends to overfit the representations of the teacher's hidden states.

Motivated by this, we propose intra-class similarity-guided feature distillation, a novel approach where we introduce a new mapping between the student and the teacher hidden representations. In fact, we match each student's sample representation with its K nearest neighbor teacher's sample representations which are within the same class to reduce the difficulty of the distillation task for the student model. Furthermore, we can look at our new mapping as a relaxation for the feature distillation task, so the student will not overfit the teacher features as detailed in Liang et al. (2023). Instead, it will focus better on the main feature extraction task while utilizing the teacher features as guidance.

In Fig. 8.1, we illustrate the key idea of our approach using a simple example. On the left side, we present the typical features matching approach where each student sample representation is mapped to its exact sample representation by the teacher. On the right side, we present our newly proposed approach where the mapping occurs between each student sample representation and its nearest teacher sample representation, from the same class. In the existing approach (Left), as sometimes the student's sample representation is very far from the teacher same sample representation, it is hard for the student to match it with its lower capacity, unlike in our proposed approach (Right) where we try to minimize the shortest distances taking advantages of the intra-class similarities.

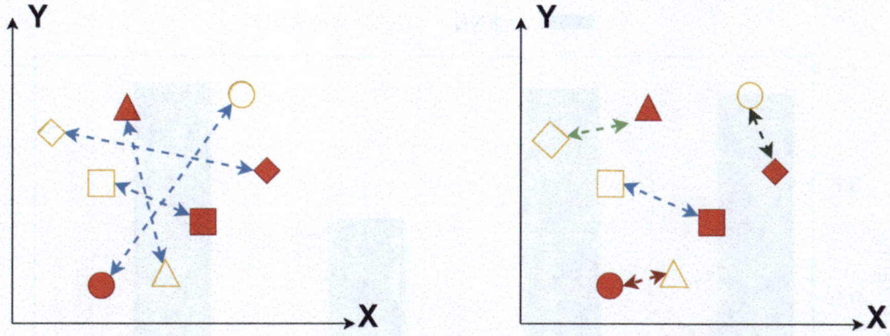

Fig. 8.1 Left: Typical feature distillation. Right: Our proposed approach. For simplicity, we set K = 1. The arrows represent the loss per sample. Red shapes (filled) represent the teacher samples representations. Yellow shapes (hollow) represent the student samples representations. The same samples are marked with the same shapes. The samples in the figure are from the same class

In our research, we distill the last hidden representation of the teacher to the student as in Tian et al. (2020) and Yang et al. (2020) where we try to group together the sample representations of the same class, revealing the intra-class similarities. This is primarily, because it is the closest to the classifier and will immediately affect the classification performance (Yang et al. 2020). We also assume that the teacher's last hidden state and the student's last hidden state have the same dimension.

8.2 Methodology

Unlike previous feature distillation work which applied a sample-wise representation alignment, we propose a KNN-based feature KD, which is a novel feature distillation method where the alignment is done between each sample representation by the student and its K nearest neighbor representations by the teacher which belong to the same class. Our approach makes the task easier for the student. Moreover, as illustrated in Fig. 8.2, the average intra-class similarity across the four GLUE benchmark datasets is higher with our method compared to the typical layer distillation technique. This highlights the effect of our approach in learning more compact class-embeddings. To empirically verify this hypothesis, we compute the intra-class cosine similarity M_{ICS} as in Eq. (8.1):

$$M_{ICS} = \frac{1}{N}\sum_{i=1}^{N}\sum_{j=1}^{c_i}\frac{< s_i \cdot s_j >}{c_i\|s_i\|_2\|s_j\|_2} \tag{8.1}$$

where N is the batch-size, s_j is the j-th sample belonging to the same class of s_i, and c_i is the total number of the samples that belong to the class of s_j in the batch of size N.

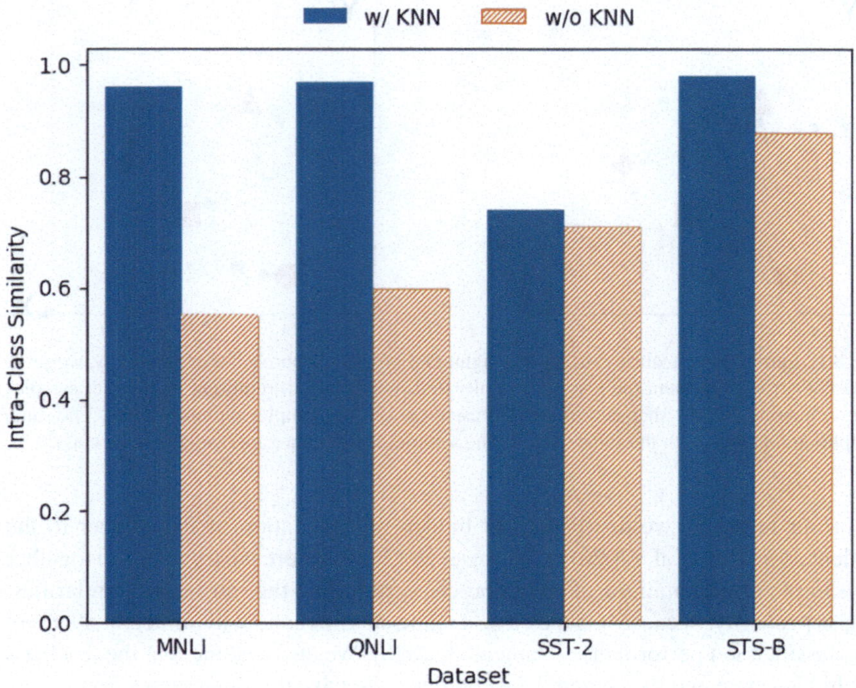

Fig. 8.2 Intra-class similarity graph. w/ KNN (solid) is the M_{ICS} metric value with our approach. w/o KNN (hatched pattern) is the M_{ICS} metric value using the exact matching for feature distillation

Typically, in a KD framework, we have the knowledgeable over-parameterized teacher modeled by f_θ. The efficient student network is modeled by $g_{\theta'}$ which has a lower number of parameters compared to the teacher $|\theta^{'}| << |\theta|$. An input batch X is fed simultaneously to f_θ and $g_{\theta'}$ to produce the last hidden representations Y_t and Y_s, respectively. Usually, to perform the feature distillation task, an MSE is computed between Y_t and Y_s (Jiao et al. 2020; Sun et al. 2019). In fact, each sample representation in Y_s is mapped to the corresponding representation in Y_t. We propose a novel mapping approach to reduce the difficulty of the task for the student. We propose to map each sample representation in Y_s to its K nearest neighbors that have the same label in Y_t. In detail, given a sample x in the input batch X where the batch X contains N samples. Its student representation r^S is with dimension n.

$r^S = g_{\theta'}(x)$. $F = \{s \mid s \in X$, and $\text{label}(x) = \text{label}(s)\}$ contains the elements in the batch that have the same label as x. $G = \{d \mid d = \sum_{j=1}^{n} \left(f_\theta(s)_j - r_j^S\right)^2, \text{and} s \in F\}$ contains the distances between each sample representation s in F by the teacher and r^S. $G_K = \{i_1, i_2, i_3, ..., i_K \mid d_{i_1} < d_{i_2} < d_{i_3} ... < d_{i_K}, \text{and} K < N\}$ contains the indices of the K-nearest points to r^S. The feature KD loss per sample is $l_{hidd}(x)$, as shown in Eq. (8.2):

$$l_{hidd}(x) = \frac{1}{n} \sum_{k \in G_K} \sum_{j=1}^{n} \left(f_\theta(s_k)_j - g_{\theta'}(x)_j\right)^2 \tag{8.2}$$

The final feature KD loss over all the batch samples is computed as in Eq. (8.3):

$$\mathcal{L}_{hidd} = \sum_{x \in X} l_{hidd}(x) \tag{8.3}$$

The final KD loss is computed as in Eq. (8.4):

$$\mathcal{L}_{KD} = \alpha_1 \mathcal{L}_{hidd} + \alpha_2 \mathcal{L}_{soft} \tag{8.4}$$

The final training loss of the student is computed as in Eq. (8.5):

$$\mathcal{L} = \mathcal{L}_{KD} + \alpha_3 \mathcal{L}_{CE} \tag{8.5}$$

where α_1, α_2, and α_3 are the contributions of the three loss components to the final training loss. $\mathcal{L}_{soft}$ is the logit distillation loss as in Sanh et al. (2019) and Jiao et al. (2020), which is the temperated KL divergence between the student logits and the teacher logits. $\mathcal{L}_{CE}$ is the cross entropy loss between the ground truth labels and the student predictions.

8.3 Experimental Results

8.3.1 Experimental Setup

Datasets In this research, we evaluate our proposed method on the validation set of seven GLUE benchmark datasets (Wang et al. 2019). The GLUE dataset is commonly used as a benchmark for KD in NLP (Zhou et al. 2021) as it is composed of several datasets for different tasks. In our evaluation, we use MNLI, QNLI, and RTE for natural language inference; SST-2 for sentiment classification; and QQP, MRPC, and STS-B for paraphrase similarity matching. The reported results follow the same format as those on the official GLUE leaderboard.

Baselines In our experiments, the teacher is a 12-layer BERT-base-uncased model, fine-tuned on each GLUE task, with 110M parameters distilled into a 6-layer BERT_6 student model with 66M parameters. The number of epochs, the sequence length, the batch size, and the learning rate are set to 5, 128, 32, and $\{1e{-}5, 3e{-}5, 5e{-}5\}$, respectively, for the teacher fine-tuning. We compare our proposed method with different state-of-the-art BERT compression approaches, including BERT-PKD (Sun et al. 2019), DistilBERT (Sanh et al. 2019), PD (Turc et al. 2019), BERT-of-Theseus (Xu et al. 2020), TinyBERT (Jiao et al. 2020), MetaDistil (Zhou et al. 2021), MiniLM v2 (Wang et al. 2021), and ReptileDistil (Ma et al. 2022).

Training settings For the baseline methods we report the same results as those in Ma et al. (2022), which are from the corresponding original paper. In our work, following Jiao et al. (2020); Ma et al. (2022), we initialize the student with the general TinyBERT$_6$ model weights. Similar to Ma et al. (2022), the sequence length, the batch size, the number of epochs, and the temperature are set to 128, 32, 5, and 5, respectively. As proposed in Sanh et al. (2019); Jiao et al. (2020), α_2 and α_3 are set to 0.5 and 0.5, respectively. Inspired by Sun et al. (2019), Zhou et al. (2021), and Ma et al. (2022), we conduct a grid search over student learning rate from $\{1e-5, 3e-5, 5e-5\}$, the K (number of nearest neighbors) from $\{1, 2, 3, 5\}$, and α_1 from $\{0.1, 0.01, 0.001\}$ and save the best model. All the experiments are repeated for four random seeds as in Sun et al. (2019) and the average is reported.

8.3.2 Results

In this section, we discuss the experimental results of our approach.

Table 8.1 Experimental results on the GLUE development sets. The numbers and the strings under each dataset name indicated the number of samples and the metrics. All student models listed have the same architecture of 6 Transformer layers with 66M parameters

	SST-2 method ACC	MRPC (67k) F1/Acc	STS-B (3.7k) pear/ spea	QQP (5.7k) F1/Acc	MNLI (364k) Acc m/ mm	QNLI (105k) Acc	RTE (2.5k) Acc
BERT$_{BASE}$ (Devlin et al. 2019) (teacher)	93.0	91.6/ 87.6	90.2/ 89.8	88.5/ 91.4	84.6/ 84.9	91.2	71.4
DistilBERT (Sanh et al. 2019)	91.3	87.5/–	–/86.9	–/88.5	82.2/–	89.2	59.9
BERT-PKD (Sun et al. 2019)	91.3	85.7/–	–/86.2	–/88.4	81.3/–	88.4	66.5
PD (Turc et al. 2019)	91.1	89.4/ 84.9	–	87.4/ 90.7	82.5/ 83.4	89.4	66.7
TinyBERT (Jiao et al. 2020)	**93.0**	90.6/ 86.3	**90.1/ 89.6**	88.0/ 91.1	84.5/ 84.5	**91.1**	73.4
BERT-of-Theseus (Xu et al. 2020)	91.5	89.0/–	–/88.7	–/89.6	82.3/–	89.5	68.2
MiniLM v2 (Wang et al. 2021)	92.4	88.9/–	–	–/91.1	84.2/–	90.8	69.4
MetaDistil (Zhou et al. 2021)	92.3	91.1/ 86.8	89.4/ 89.1	**88.1/ 91.0**	83.5/ 83.8	90.4	72.1
ReptileDistil (Ma et al. 2022)	92.2	91.6/ 87.7	89.5/ 89.3	87.6/ 90.1	83.7/ 83.7	90.5	75.3
Ours	92.5	**92.5/ 89.64**	89.7/ 89.5	87.7/ 90.9	**84.5/ 84.5**	90.8	**75.8**

Table 8.2 Experimental results on the RTE and MRPC development sets. The reported results are averaged over four random seeds

Approach	MRPC (%)	RTE (%)
Teacher	93.05/90.19	87.00
Student	92.48/89.52	84.83
Our model	93.29/90.68	86.19

As shown in Table 8.1, our proposed approach outperforms all state-of-the-art methods on three datasets, i.e., MRPC, MNLI, and RTE. While distilling knowledge from a static teacher, ours outperforms both KD state-of-the-art MetaDistil and ReptileDistil, where the teacher is dynamic, on most of the datasets. While we distill the knowledge only from the last hidden representation of the teacher, ours outperforms BERT-PKD on all the datasets, which distills several hidden representations from the teacher to the student. It is also worth mentioning that although Wang et al. (2023) showed that the attention distillation is the best-performing objective, ours outperforms MiniLM v2, which distills the attention on all the datasets, and TinyBERT, which distills the attention, all the hidden states, and the logits, on three datasets.

Evaluation with DEBERTA

We also conducted an additional small experiment comparing our approach with larger DEBERTA variants (He et al. 2021). In this experiment, the teacher is DEBERTA-Xlarge, which has 750M parameters and the student is DEBERTA-large with 400M parameters. The RTE and MRPC scores of our approach along with the standard student fine-tuning are presented in Table 8.2. As can be seen, the proposed approach still helps in this case and consistently yields superior performance.

8.4 Evaluation on HateSpeech

8.4.1 Experimental Setup

8.4.1.1 Dataset

In this section, we evaluate our proposed KD approach on the hate-speech detection task, using the OffensEval (Zampieri et al. 2019), AbusEval (Caselli et al. 2020) and HatEval (Basile et al. 2019) English hate-speech datasets.

OffensEval2019 is composed of 14,100 tweets that are binary annotated, as either offensive or non-offensive. A tweet is considered offensive if "*it contains any form of non-acceptable language (profanity) or a targeted offense, which can be veiled or direct*" (Zampieri et al. 2019). The dataset is split into 13,240 training examples with 4,400 offensive tweets (i.e., the positive class) and 860 test examples with 240 offensive tweets.

Abuseval is OffensEval2019 annotated for abusive versus non-abusive language. A tweet is labelled as abusive if "*it contains hurtful language that a speaker uses to insult or offend an individual or a group of individuals based on personal qualities, appearance, social status, opinions, statements, or actions*" (Caselli et al. 2020). It has the same training size and test size as OffensEval2019.

HatEval is composed of 13,000 tweets that are binary annotated hateful vs non-hateful against migrants and women. A tweet is considered hateful if "*any communication that disparages a person or a group on the basis of some characteristic such as race, colour, ethnicity, gender, sexual orientation, nationality, religion, or other characteristics*" (Basile et al. 2019). The dataset is split into 10,000 training examples with 4,165 hateful messages and 3,000 test examples with 1,252 hateful messages.

8.4.1.2 Training Settings

In this distillation framework, the teacher is the HateBERT model (Caselli et al. 2021), fine-tuned on each of the datasets. HateBERT is a retrained 12-layer BERT for English hate-speech detection. The model was trained on RAL-E (Caselli et al. 2021), a large dataset collected from Reddit hateful comments. The students are a 6-layer BERT model initialized with the first 6 layers of the pre-trained HateBERT and a 3-layer BERT model initialized with the first 3 layers of the pre-trained HateBERT. In this evaluation, we compare the students' performance to the BERT-base model and the teacher, as in Caselli et al. (2021).

For the teacher and the BERT-base fine-tuning, we follow a similar experimental setup described in Caselli et al. (2021). The learning rate, batch size, max sequence length, and number of epochs are set to $1e-5$, 32, 100, and 5, respectively. For the distillation, α_2 and α_3 are set to 0.5 and 0.5, respectively, and the student learning rate is set to $5e-5$. The batch size is 32, the temperature is 2, and the number of epochs is 3.

Following the experimental setup in Sect. 8.3, K ranges from $\{1, 2, 3, 5\}$ and α_1 is set to 0.01. All the experiments are repeated for four random seeds as in Sun et al. (2019) and both the average and standard deviation are reported.

8.4.2 Results and Discussion

In this section, we discuss the evaluation metrics used and the results obtained. In these experiments, Macro average F1-score (Macro-F1) and the F1-score of the positive class (Pos .class-F1) are the reported evaluation metrics following Caselli et al. (2021).

As illustrated in Table 8.3, while our student model (i.e., the Distilled-HateBERT$_6$) has only half the number of the parameters of the teacher and BERT$_{12}$ models, on the AbusEval dataset, it outperforms BERT$_{12}$ by 1.7% as Macro-F1 and 2.6% as Pos

Table 8.3 Experimental results on the OffensEval, HatEval, and AbusEval test sets. Each experiment is repeated for four random seeds and the average and the standard deviation are reported

Dataset	Model	Macro-F1	Pos .class-F1
OffensEval 2019	HateBERT_{12} (Caselli et al. 2021)(Teacher)	0.780	0.682
OffensEval 2019	BERT_{12} (Devlin et al. 2019)	0.775±0.007	**0.682±0.006**
	Distilled-HateBERT_{6}(Student)	**0.782±0.001**	0.675±0.005
	Distilled-HateBERT_{3}(Student)	0.761±0.003	0.653±0.004
HatEval	HateBERT_{12} (Caselli et al. 2021)(Teacher)	0.531	0.641
HatEval	BERT_{12} (Devlin et al. 2019)	0.490±0.021	**0.632±0.003**
	Distilled-HateBERT_{6}(Student)	**0.496±0.017**	0.631±0.001
	Distilled-HateBERT_{3}(Student)	0.451±0.015	0.620±0.001
AbusEval	HateBERT_{12} (Caselli et al. 2021)(Teacher)	0.754	0.598
AbusEval	BERT_{12} (Devlin et al. 2019)	0.728±0.009	0.554±0.017
	Distilled-HateBERT_{6}(Student)	**0.745±0.008**	**0.580±0.018**
	Distilled-HateBERT_{3}(Student)	0.710±0.008	0.516±0.014

.class-F1. Moreover, its performance drops only by 0.9% for Macro-F1 and 1.8% for Pos .class-F1 compared to the teacher.

On OffensEval2019, Distilled-HateBERT_6 has comparable results to BERT_{12}. It is worth mentioning that it also could slightly exceed the performance of the teacher on the Macro-F1 metric. This behaviour is commonly observed in Furlanello et al. (2018) and Stanton et al. (2021). On HatEval, Distilled-HateBERT_6 slightly outperforms BERT_{12}. Furthermore, the standard deviations of the Distilled-HateBERT_6 student model are lower than those of the BERT model in most cases, which demonstrates the stability and consistency of our approach.

Additionally, we distilled the teacher HateBERT_{12} model to a 3-layer BERT model that is initialized by the first 3 layers of HateBERT_{12}. As can be seen from Table 8.3, the resultant student model (i.e., Distilled-HateBERT_3) has just one third of the parameters of the teacher and BERT_{12} models, but its performance did not drop dramatically in comparison.

8.5 Conclusion

In this research, we introduced a new mapping between the hidden representations of the teacher and the student. In fact, each sample representation by the student is mapped to its K nearest neighbor representations by the teacher. Our approach makes the task easier for the student and helps it to learn more compact sample embeddings. Empirical results on the GLUE benchmark dataset showed the effectiveness of our proposed method in comparison to competitive approaches for model compression with KD such as ReptileDistil and MetaDistil. In addition to being simple and easy to implement, our novel method outperformed existing KD approaches on three GLUE tasks. Furthermore, we evaluated our distillation approach on three hate-speech detection tasks, where we distilled the 12-layer HateBERT model to a smaller efficient one (i.e., Distilled-HateBERT with 6 Transformer layers) without significantly compromising accuracy. Future work will include exploring adding a projector to dispose of the requirement that the student and the teacher must have the same final hidden state dimension.

Acknowledgements The project on which this report is based was funded by the German Federal Ministry of Education and Research (BMBF) under the funding code 01IS20049. The author is responsible for the content of this publication.

SPONSORED BY THE

Federal Ministry
of Education
and Research

References

Basile V, Bosco C, Fersini E, Nozza D, Patti V, Rangel Pardo FM, Rosso P, Sanguinetti M (2019) SemEval-2019 task 5: Multilingual detection of hate speech against immigrants and women in Twitter. In: Proceedings of the 13th international workshop on semantic evaluation, Minneapolis, Minnesota, USA, pp 54–63. https://aclanthology.org/S19-2007

Caselli T, Basile V, Mitrović J, Kartoziya I, Granitzer M (2020) I feel offended, don't be abusive! implicit/explicit messages in offensive and abusive language. In: Proceedings of the twelfth language resources and evaluation conference, Marseille, France, pp 6193–6202. https://aclanthology.org/2020.lrec-1.760

Caselli T, Basile V, Mitrović J, Granitzer M (2021) HateBERT: Retraining BERT for abusive language detection in English. In: Proceedings of the 5th workshop on online Abuse and Harms (WOAH 2021), Online, pp 17–25. https://aclanthology.org/2021.woah-1.3

Chen Y, Wang S, Liu J, Xu X, de Hoog F, Huang Z (2022) Improved feature distillation via projector ensemble. Adv Neural Inf Process Syst 35:12084–12095

Devlin J, Chang MW, Lee K, Toutanova K (2019) BERT: pre-training of deep bidirectional transformers for language understanding. In: Proceedings of the 2019 conference of the North American chapter of the association for computational linguistics (ACL): human language technologies, vol 1 (Long and short papers), pp 4171–4186

Furlanello T, Lipton Z, Tschannen M, Itti L, Anandkumar A (2018) Born again neural networks. In: International conference on machine learning, PMLR, pp 1607–1616

He P, Liu X, Gao J, Chen W (2021) Deberta: decoding-enhanced bert with disentangled attention. In: International conference on learning representations (ICLR)

Hinton G, Vinyals O, Dean J (2015) Distilling the knowledge in a neural network. arXiv preprint arXiv:1503.02531

Jiao X, Yin Y, Shang L, Jiang X, Chen X, Li L, Wang F, Liu Q (2020) TinyBERT: distilling BERT for natural language understanding. In: Findings of the association for computational linguistics:

EMNLP 2020, Association for Computational Linguistics, Online, pp 4163–4174. https://aclanthology.org/2020.findings-emnlp.372

Kovaleva O, Romanov A, Rogers A, Rumshisky A (2019) Revealing the dark secrets of BERT. In: Proceedings of the 2019 conference on empirical methods in natural language processing and the 9th international joint conference on natural language processing (EMNLP-IJCNLP), Hong Kong, China, pp 4365–4374. https://aclanthology.org/D19-1445

Liang C, Zuo S, Zhang Q, He P, Chen W, Zhao T (2023) Less is more: Task-aware layer-wise distillation for language model compression. In: International conference on machine learning, PMLR, pp 20852–20867

Ma X, Wang J, Yu LC, Zhang X (2022) Knowledge distillation with reptile meta-learning for pretrained language model compression. In: Proceedings of the 29th international conference on computational linguistics, pp 4907–4917

Passban P, Wu Y, Rezagholizadeh M, Liu Q (2021) ALP-KD: Attention-based layer projection for knowledge distillation. Proceedings of the AAAI conference on artificial intelligence 35:13657–13665

Romero A, Ballas N, Kahou SE, Chassang A, Gatta C, Bengio Y (2014) Fitnets: hints for thin deep nets. In: International conference on learning representations

Saadi K, Mitrović J, Granitzer M (2023) Learn from one specialized sub-teacher: one-to-one mapping for feature-based knowledge distillation. In: ICML 2023 workshop neural compression: from information theory to applications

Sanh V, Debut L, Chaumond J, Wolf T (2019) Distilbert, a distilled version of bert: smaller, faster, cheaper and lighter. arXiv preprint arXiv:1910.01108

Stanton S, Izmailov P, Kirichenko P, Alemi AA, Wilson AG (2021) Does knowledge distillation really work? Adv Neural Inf Process Syst 34:6906–6919

Sun S, Cheng Y, Gan Z, Liu J (2019) Patient knowledge distillation for bert model compression. In: Proceedings of the 2019 conference on empirical methods in natural language processing and the 9th international joint conference on natural language processing (EMNLP-IJCNLP), pp 4323–4332

Tian Y, Krishnan D, Isola P (2020) Contrastive representation distillation

Turc I, Chang MW, Lee K, Toutanova K (2019) Well-read students learn better: on the importance of pre-training compact models. arXiv preprint arXiv:1908.08962

Wang A, Singh A, Michael J, Hill F, Levy O, Bowman S (2019) GLUE: a multi-task benchmark and analysis platform for natural language understanding. In: International conference on learning representations (ICLR)

Wang W, Bao H, Huang S, Dong L, Wei F (2021) MiniLMv2: multi-head self-attention relation distillation for compressing pretrained transformers. In: Findings of the association for computational linguistics: ACL-IJCNLP 2021. Association for Computational Linguistics, pp 2140–2151. https://aclanthology.org/2021.findings-acl.188

Wang X, Weissweiler L, Schütze H, Plank B (2023) How to distill your BERT: an empirical study on the impact of weight initialisation and distillation objectives. In: Proceedings of the 61st annual meeting of the association for computational linguistics (Volume 2: Short papers), Toronto, Canada, pp 1843–1852. https://aclanthology.org/2023.acl-short.157

Wu Y, Passban P, Rezagholizadeh M, Liu Q (2020) Why skip if you can combine: a simple knowledge distillation technique for intermediate layers. In: Proceedings of the 2020 conference on empirical methods in natural language processing (EMNLP), Online, pp 1016–1021. https://aclanthology.org/2020.emnlp-main.74

Xu C, Zhou W, Ge T, Wei F, Zhou M (2020) BERT-of-theseus: compressing BERT by progressive module replacing. In: Proceedings of the 2020 conference on empirical methods in natural language processing (EMNLP). Association for Computational Linguistics, pp 7859–7869. https://aclanthology.org/2020.emnlp-main.633

Yang J, Martinez B, Bulat A, Tzimiropoulos G (2020) Knowledge distillation via softmax regression representation learning. In: International conference on learning representations (ICLR)

Zampieri M, Malmasi S, Nakov P, Rosenthal S, Farra N, Kumar R (2019) SemEval-2019 task 6: Identifying and categorizing offensive language in social media (OffensEval). In: Proceedings of the 13th international workshop on semantic evaluation. Association for Computational Linguistics, Minneapolis, Minnesota, USA, pp 75–86. https://aclanthology.org/S19-2010

Zhou W, Xu C, McAuley J (2021) Bert learns to teach: knowledge distillation with meta learning. In: Annual meeting of the association for computational linguistics. https://api.semanticscholar.org/CorpusID:237250417

Part V
Efficiency Techniques for Smaller Scales

Chapter 9
On the Use of Cross-Attentive Fusion Techniques for Audio-Visual Speaker Verification

Jahangir Alam and R. Gnana Praveen

Abstract Audio-Visual (A-V) speaker verification has recently been gaining a lot of attention owing to the closely associated audio and visual cues. Though existing approaches based on A-V fusion showed improvement over unimodal systems, its potential for speaker verification is not fully exploited. In this contribution, we investigate the prospect of effectively capturing the synergic relationships across audio and visual modalities, which can play a vital role to significantly boost the performance of multimodal fusion over unimodal systems. More specifically, we present a comparative study of three variants of Cross-Attentive frameworks, namely Joint Cross-Attention (JCA), Recursive JCA (RJCA) and Cross-Modal Transformer (CMT), for multimodal fusion that can efficiently capture the intra-modal and/or inter-modal associations for A-V speaker verification. We carry out extensive experiments on the Voxceleb1 dataset to rigorously evaluate and compare the cross-attention-based models. Results indicate that effectively capturing intra- and/or inter-modal relationships across audio and visual modalities can significantly improve the performance of the A-V speaker verification system.

9.1 Introduction

Speaker Verification (SV) is the process of verifying the identity of an individual based on unique voice characteristics. It has a wide range of applications such as access control, unlocking devices, authorizing transactions, forensics, commercial, and law enforcement applications (Hansen and Hasan 2015). The task of SV has been predominantly explored using faces (Wang and Deng 2021) and speech (Tu et al. 2022) signals independently. With the advancement of deep learning models, both face- and speech-based methods have individually achieved remarkable success (Tu et al. 2022). However, relying on single modalities may often deteriorate the performance of the system when face- or speech-based signals are degraded by extreme

J. Alam (✉) · R. G. Praveen
Computer Research Institute of Montreal (CRIM), Montreal (Quebec), Canada
e-mail: jahangir.alam@crim.ca

© The Author(s), under exclusive license to Springer Nature Switzerland AG 2025
P. Passban et al. (eds.), *Enhancing LLM Performance*, Machine Translation: Technologies and Applications 7, https://doi.org/10.1007/978-3-031-85747-8_9

background noise or intra-variations such as pose, low illumination, or manner of speaking. Therefore, the fusion of facial and vocal data, which often reliably complements each other, is gaining momentum, enhancing security and robustness under adverse conditions (Tao et al. 2023). For instance, when speech modality is corrupted, we can rely on face to verify the identity of a person and vice-versa. Most of the existing audio-visual (A-V) fusion approaches for SV have focused on score-level fusion (Alam et al. 2020; Seyed et al. 2020) or early feature-level fusion (Chen et al. 2020; Qian et al. 2021). Although these methods have improved fusion performance over unimodal systems, they fail to leverage the rich complementary inter-modal relationships between the audio and visual modalities.

In recent years, attention-based models have been explored to efficiently capture complementary inter-modal associations across faces and voices (Hörmann et al. 2020; Sun et al. 2023). Most existing attention-based models attempted to leverage the intra- and inter-modal relationships in a decoupled fashion. Another approach involves mitigating the impact of noisy modalities by leveraging their complementary relationships (Shon et al. 2019; Tao et al. 2023). To effectively fuse audio and visual modalities, it is very important to adeptly capture intra- and/or inter-modal relationships. Intra-modal relationships offer valuable insights into the temporal dynamics of videos, while inter-modal relationships provide substantial information regarding the complementarity between modalities. In this work, for improving the performance of multimodal fusion, we aim to investigate the prospect of effectively capturing the intra- and/or inter-modal relationships across audio and visual modalities, and provide a comparative study of multiple variants of cross-attention mechanisms. More explicitly, we focus on three variants of Cross-Attentive frameworks for multimodal fusion, namely, Joint Cross-Attention (JCA), Recursive JCA (RJCA) and Cross-Modal Transformer (CMT), that can adeptly cover the intra- and/or inter-modal associations for A-V speaker verification.

The major contributions of this chapter can be summarized as follows:

- Inspired by the success of multimodal transformers in various applications, we adopt a CMT model, which employs Transformers in the framework of cross-attention for A-V speaker verification.
- We investigate multiple variants of cross-attention-based mechanisms (JCA, RJCA and CMT) and provide a comparative study of the impact of these mechanisms for A-V speaker verification.
- Extensive experiments are conducted on the voxceleb1 dataset to demonstrate the productiveness of the JCA, RJCA and CMT models for multimodal SV tasks.

9.2 Related Work

The close association between face and voice has attracted much attention for the task of cross-modal biometric matching by projecting the features of individual modalities to a common representation space (Nagrani et al. 2018b, a). Sarı et al. (2021)

explored a multi-view approach by transforming the individual feature representations into a common latent space and using a shared classifier for both modalities for SV. Chen et al. (2023) leveraged complementary information as a means of supervision to obtain robust A-V feature representations using a co-meta-learning paradigm in a self-supervised learning framework. Tao et al. (2023) also explored the complementary relationship between audio and visual modalities to clean noisy samples, where consistency across audio and visual modalities is used to discriminate easy from hard samples. Another line of investigation is to mitigate the impact of noisy modalities by leveraging complementary relationships. Shon et al. (2019) proposed an attention mechanism to assign higher attention scores to the modality exhibiting higher discrimination by benefiting from complementarities across audio and visual modalities. Hörmann et al. (2020) further extended the idea of Shon et al. (2019) by introducing feature fusion of audio and visual modalities at intermediate layers to improve the quality of feature representations. Chen et al. (2020) explored the prospect of obtaining robust feature representations by investigating various fusion strategies at the embedding level and achieved best performance using gating-based fusion. They also implemented data augmentation strategies to address issues with extremely corrupted or missing modalities.

None of the above approaches leverage cross-modal interactions to effectively capture the rich inter-modal relationships. Cross-modal attention has been successfully explored in several applications such as weakly supervised action localization (Lee et al. 2021), event localization (Duan et al. 2021), and emotion recognition (Praveen et al. 2021). Recently, Mocanu and Tapu (2022) also explored cross-attention (CA) based on cross-correlations across audio and visual modalities to effectively capture the complementary relationships between them. Liu et al. (2023) explored cross-modal attention by deploying cross-modal boosters in a pseudo-Siamese structure to model one modality by exploiting knowledge from another modality. However, these approaches focus only on inter-modal relationships (Mocanu and Tapu 2022) or capture intra- and inter-modal relationships in a decoupled fashion (Liu et al. 2023). Rajasekhar and Alam (2023) explored a joint cross-attentional framework to jointly capture the intra and inter-modal relationships and showed better performance of SV. Recursive attention has also been previously explored successfully for emotion recognition (Praveen et al. 2023b) and event localization (Duan et al. 2021). In this work, we perform an analytical study on three cross-attention variants, namely JCA, RJCA, and cross-modal Transformers (CMT), and further explore the prospects of CMT-based multimodal fusion for A-V person verification.

9.2.1 Applicability to LLMs

This chapter focuses on developing fusion models to effectively combine information from audio and visual modalities to enhance person verification performance. These fusion models operate on deep features extracted from pre-trained audio and visual networks, aiming to leverage the complementary nature of the two modalities. Compared to unimodal systems, multimodal systems can be expected to offer improved robustness and security, especially in challenging conditions such as missing or noisy modality and potential deepfake attacks. While this research doesn't directly address techniques for LLMs, the concepts and methodologies explored could potentially be applied to the development of multimodal LLMs. By integrating information from multiple modalities, multimodal LLMs could offer more comprehensive and contextually rich understanding, which could be beneficial for various natural language understanding and generation tasks. Moreover, the idea of leveraging fusion models to combine information from different sources could potentially enhance the performance and personalization capabilities of LLMs in diverse applications.

9.3 Cross-Attention Mechanisms

Attention mechanisms have emerged as a powerful tool in deep learning, enabling models to focus on relevant information and ignore less important parts, thereby capturing dependencies and relationships within the input. Since its introduction, attention mechanisms have been successfully applied in different domains such as natural language processing (Kretov 2020), computer vision (Guo et al. 2022), and speech processing applications (Alam 2023).

In contrast, in the context of multimodal fusion, a CA mechanism is an extension to the idea of attention mechanisms, allowing the model to attend to relevant information from multiple modalities simultaneously instead of attending to only parts of a single modality. Cross-attention-based multimodal fusion has been successfully applied to various tasks, including multimodal sentiment analysis (Jiang and Ji 2022), image captioning (Zhao et al. 2019), visual question answering (Zhang and Wu 2022), speaker recognition (Liu et al. 2023; Mocanu and Tapu 2022) and so on.

Since multimodal data convey more diverse information than unimodal data, effectively leveraging the intra-modal and/or inter-modal complementary relationships among both audio and visual modalities plays a crucial role in efficient audio-visual fusion. In this section, we introduce three variants of cross-attentive multimodal fusion for A-V speaker verification.

9.3.1 Problem Formulation

For an input video sub-sequence S, L non-overlapping video segments are uniformly sampled, and the corresponding deep feature vectors are extracted using the pre-trained models of audio and visual modalities. Let $\mathbf{Z_a}$ and $\mathbf{Z_v}$ denote the deep feature vectors of audio and visual modalities, respectively, for the given input video sub-sequence $\boldsymbol{S}$ of fixed size, which is expressed in (9.1) and (9.2):

$$\mathbf{Z_a} = \{z_{\mathbf{a}}^1, z_{\mathbf{a}}^2, ..., z_{\mathbf{a}}^L\} \in \mathbb{R}^{d_a \times L} \tag{9.1}$$

$$\mathbf{Z_v} = \{z_{\mathbf{v}}^1, z_{\mathbf{v}}^2, ..., z_{\mathbf{v}}^L\} \in \mathbb{R}^{d_v \times L} \tag{9.2}$$

where d_a and d_v represent the dimensions of the audio and visual feature vectors, respectively, and $z_{\mathbf{a}}^l$ and $z_{\mathbf{v}}^l$ denote the audio and visual feature vectors of the video segments, respectively, for $l = 1, 2, ..., L$ segments.

The objective of the problem is to estimate the speaker verification model $f : \boldsymbol{Z} \rightarrow \boldsymbol{Y}$ from the training data $\mathbf{Z}$, where $\mathbf{Z}$ denotes the set of audio and visual feature vectors of the input video segments and $\boldsymbol{Y}$ represents the speaker identity of the corresponding video sub-sequence S.

It has been observed that the performance of unified multimodal training may decline over that of individual modalities due to the differences in learning dynamics and noise topologies (Wang et al. 2020). Therefore, we have obtained deep feature vectors for the individual audio and visual modalities independently, which are then fed to the cross-attentional module for A-V fusion.

9.3.2 Joint Cross-Attention (JCA)

In this section, we present Joint Cross-Attention (JCA)-based multimodal fusion to simultaneously encode intra-modal and inter-modal relationships in a joint framework. The JCA approach was initially proposed in Praveen et al. (2022, 2023a) for the multimodal dimensional emotion recognition task, but was later adopted in Rajasekhar and Alam (2023) for SV. Specifically, the joint A-V feature representation obtained by concatenating the audio and visual features is fed to the fusion module along with the feature representations of the individual modalities. By deploying the joint representation, features of each modality attend to themselves, as well as other modalities, thereby simultaneously capturing the semantic inter-modal and intra-modal relationships among audio and visual modalities. Leveraging the joint representation also helps to reduce heterogeneity in the audio and visual modalities, which further improves SV performance. A schematic diagram of the JCA model is depicted in Fig. 9.1, where both $\mathbf{Z_a}$ and $\mathbf{Z_v}$ are passed through a bidirectional LSTM (BLSTM) module with residual embedding so as to encode the temporal dynamics of the segments of the sequence of audio and visual feature vectors, respectively. Let $\boldsymbol{X}_\mathbf{a}$ and $\boldsymbol{X}_\mathbf{v}$ denote the deep feature vectors of audio and visual modalities, respectively, which is placed after BLSTM module, as explained in (9.3) and (9.4):

$$\boldsymbol{X}_\mathbf{a} = \{\boldsymbol{x}_\mathbf{a}^1, \boldsymbol{x}_\mathbf{a}^2, ..., \boldsymbol{x}_\mathbf{a}^L\} \in \mathbb{R}^{d_a \times L} \tag{9.3}$$

$$\boldsymbol{X}_\mathbf{v} = \{\boldsymbol{x}_\mathbf{v}^1, \boldsymbol{x}_\mathbf{v}^2, ..., \boldsymbol{x}_\mathbf{v}^L\} \in \mathbb{R}^{d_v \times L} \tag{9.4}$$

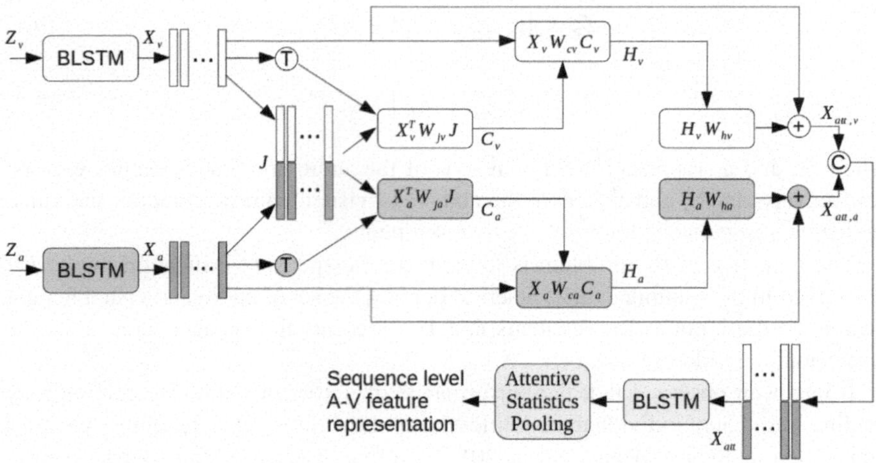

Fig. 9.1 Components of the JCA model for A-V fusion for SV. Here, modules (C) and (T) indicate the concatenation and transpose operation, respectively

The joint representation $\boldsymbol{J}$ is obtained by concatenating the audio and visual feature vectors as $\boldsymbol{J} = [\boldsymbol{X}_{\mathbf{a}}; \boldsymbol{X}_{\mathbf{v}}] \in \mathbb{R}^{d \times L}$ where $d = d_a + d_v$ denotes the feature dimension of the concatenated features. The concatenated A-V feature representations ($\boldsymbol{J}$) of the given video sub-sequence ($\boldsymbol{S}$) are now employed in the CA framework to attend to the individual modalities. This helps to incorporate both intra- and inter-modal relationships in obtaining the attention weights of audio and visual modalities. Now the correlation across the joint feature representation and the individual modalities are obtained as joint cross-correlation matrices, which are computed as in Eq. (9.5):

$$\boldsymbol{C}_{\mathbf{a}} = \tanh\left(\frac{\boldsymbol{X}_{\mathbf{a}}^{\top} \boldsymbol{W}_{\mathbf{ja}} \boldsymbol{J}}{\sqrt{d}}\right) \quad \text{and} \quad \boldsymbol{C}_{\mathbf{v}} = \tanh\left(\frac{\boldsymbol{X}_{\mathbf{v}}^{\top} \boldsymbol{W}_{\mathbf{jv}} \boldsymbol{J}}{\sqrt{d}}\right) \tag{9.5}$$

where $\boldsymbol{W}_{\mathbf{ja}} \in \mathbb{R}^{d_a \times d}$ and $\boldsymbol{W}_{\mathbf{jv}} \in \mathbb{R}^{d_v \times d}$ represent the learnable weight matrices of the audio and visual modalities, respectively. The joint correlation matrices $\boldsymbol{C}_{\mathbf{a}}$ and $\boldsymbol{C}_{\mathbf{v}}$ for audio and visual modalities help to obtain the attention weights based on the semantic relevance of both across and within the modalities. The higher the correlation coefficient, the higher the correlation across the corresponding samples within the same modality as well as across modalities. Now the joint cross-correlation matrices are used to obtain the attention maps of audio and visual modalities, as shown in Eq. (9.6):

$$\boldsymbol{H}_{\mathbf{a}} = ReLU(\boldsymbol{X}_{\mathbf{a}} \boldsymbol{W}_{\mathbf{ca}} \boldsymbol{C}_{\mathbf{a}}) \quad \text{and} \quad \boldsymbol{H}_{\mathbf{v}} = ReLU(\boldsymbol{X}_{\mathbf{v}} \boldsymbol{W}_{\mathbf{cv}} \boldsymbol{C}_{\mathbf{v}}) \tag{9.6}$$

where $\boldsymbol{W}_{\mathbf{ca}} \in \mathbb{R}^{L \times L}, \boldsymbol{W}_{\mathbf{cv}} \in \mathbb{R}^{L \times L}$ denote learnable matrices of audio and visual modalities respectively. These attention maps are used to obtain the attended features of the audio and visual modalities, computed as in Eq. (9.7):

$$\boldsymbol{X}_{\mathbf{att,a}} = \boldsymbol{H}_{\mathbf{a}}\boldsymbol{W}_{\mathbf{ha}} + \boldsymbol{X}_{\mathbf{a}} \quad \text{and} \quad \boldsymbol{X}_{\mathbf{att,v}} = \boldsymbol{H}_{\mathbf{v}}\boldsymbol{W}_{\mathbf{hv}} + \boldsymbol{X}_{\mathbf{v}} \tag{9.7}$$

where $\boldsymbol{W}_{\mathbf{ha}} \in \mathbb{R}^{L\times L}$ and $\boldsymbol{W}_{\mathbf{hv}} \in \mathbb{R}^{L\times L}$ denote the learnable weight matrices for audio and visual modalities, respectively. The attended audio and visual features $\boldsymbol{X}_{\mathbf{att,a}}$ and $\boldsymbol{X}_{\mathbf{att,v}}$ are further concatenated to obtain the A-V feature representation, which is given by $\widehat{X}$, as in Eq. (9.8):

$$\widehat{X} = [\boldsymbol{X}_{\mathbf{att,a}}; \boldsymbol{X}_{\mathbf{att,v}}] \tag{9.8}$$

As shown in Fig. 9.1, the attended A-V feature vectors are fed to the BLSTM in order to capture the temporal dynamics of the attended joint A-V feature representations. The segment-level A-V feature representations are in turn fed to the attentive statistics pooling (ASP) (Okabe et al. 2018) to obtain the sub-sequence or utterance-level representation of the A-V feature vectors. Finally, the embeddings of the final A-V feature representations are used to obtain the scores, where the additive angular margin Softmax (AAMSoftmax) (Deng et al. 2018) loss function is used to optimize the parameters of the fusion model and the ASP module.

9.3.3 *Recursive Joint Cross-Attention (RJCA)*

In this section, we present the extended version of the JCA model proposed in Praveen and Alam (2024a) by including a recursive mechanism to obtain more refined feature representations. A schematic diagram of this recursive JCA model is shown in Fig. 9.2, and we call this fusion model "Recursive Joint Cross-Attention" (RJCA). The main idea behind deploying the joint feature representation in the CA framework in a recursive fashion is to simultaneously boost intra- and inter-modelling of A-V relationships. Unlike the JCA approach, the recursive JCA does not employ any BLSTM module, so the input features to the RJCA fusion model are $\boldsymbol{X}_{\mathbf{a}} = \boldsymbol{Z}_{\mathbf{a}}$ and $\boldsymbol{X}_{\mathbf{v}} = \boldsymbol{Z}_{\mathbf{v}}$.

The JCA and RJCA (as illustrated in Figs. 9.1 and 9.2) approaches share similar processing stages. The only difference is that, in RJCA, an additional recursion step, as indicated by the green line in Fig. 9.2, is appended. More specifically, in RJCA fusion approach, in order to attain more refined feature representations, the attended features are again fed as input, as shown in Eq. (9.9):

$$\boldsymbol{X}^{(t)}_{\mathbf{att,a}} = \boldsymbol{H}^{(t)}_{\mathbf{a}}\boldsymbol{W}^{(t)}_{\mathbf{ha}} + \boldsymbol{X}^{(t-1)}_{\mathbf{a}} \quad \text{and} \quad \boldsymbol{X}^{(t)}_{\mathbf{att,v}} = \boldsymbol{H}^{(t)}_{\mathbf{v}}\boldsymbol{W}^{(t)}_{\mathbf{hv}} + \boldsymbol{X}^{(t-1)}_{\mathbf{v}} \tag{9.9}$$

where $\boldsymbol{W}^{(t)}_{\mathbf{ha}} \in \mathbb{R}^{L\times L}$ and $\boldsymbol{W}^{(t)}_{\mathbf{hv}} \in \mathbb{R}^{L\times L}$ denote the learnable weight matrices of the tth iteration for audio and visual modalities, respectively. Finally the attended features $\boldsymbol{X}^{(t)}_{\mathbf{att,a}}$ and $\boldsymbol{X}^{(t)}_{\mathbf{att,v}}$ obtained from the recursive fusion model are concatenated and fed to the attentive statistics pooling (ASP) module to obtain sub-sequence or utterance-level representations. The utterance-level A-V feature representations are used to

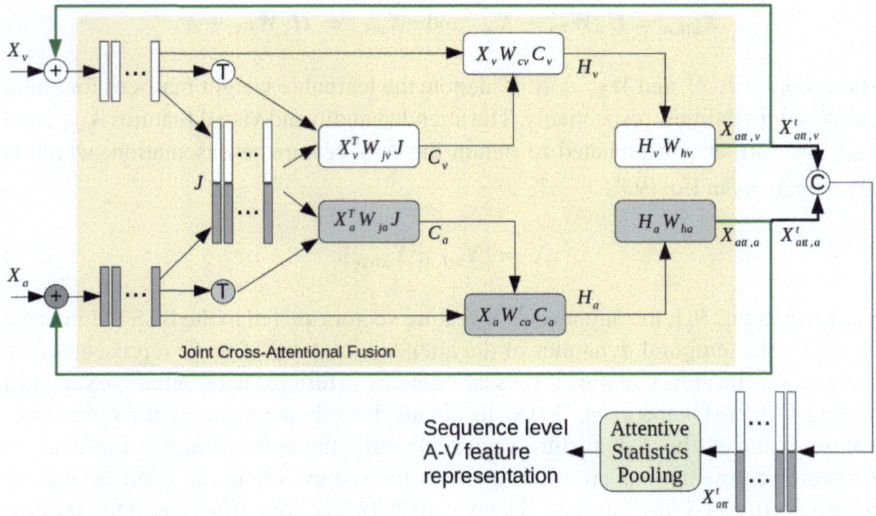

Fig. 9.2 Components of the RJCA model for A-V fusion. Here, modules (C) and (T) indicate the concatenation and transpose operation, respectively

obtain the scores, where the AAMSoftmax loss function (Deng et al. 2018) is used to optimize the parameters of the fusion model and the ASP module.

9.3.4 Cross-Modal Transformers (CMT)

In this section, we introduce our Cross-Modal Transformer (CMT)-based multimodal fusion approach (Praveen and Alam 2024b) for the A-V SV task. A schematic diagram of the various steps in CMT is shown in Fig. 9.3. As described in Sect. 9.3.1, deep features $\mathbf{Z_a}$ and $\mathbf{Z_v}$ of audio and visual modalities are extracted via the pre-trained audio (e.g., ECAPA-TDNN) and visual (e.g., Resnet-18) models, respectively. Similar to the JCA framework, the extracted deep features $\mathbf{Z_a}$ and $\mathbf{Z_v}$ are then passed individually through a BLSTM module with residual embedding for encoding the temporal dynamics of the segments of the sequence of the audio and visual feature vectors, respectively. The outputs of the BLSTM module are termed $\boldsymbol{X_a}$ and $\boldsymbol{X_v}$.

This approach consists of two Transformer encoders, one for each modality. Typically, each encoder comprises multiple self-attention layers, each producing three types of representations—Keys (K), Values (V), and Queries (Q)—from the outputs of previous layer. The K and Q matrices are used to compute the attention by computing a dot product between them, as given by Eq. (9.10):

$$Y = Softmax(KQ^T)V \tag{9.10}$$

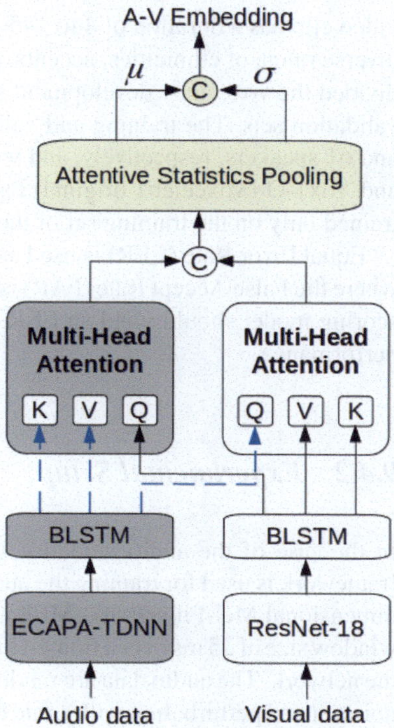

Fig. 9.3 Illustration of the proposed CMT model. It employs the ECAPA-TDNN and ResNet-18 architectures as the audio-only and visual-only encoding networks, respectively. The module (C) represents the concatenation operation, and μ and σ indicate the weighted mean and standard deviation, respectively

Based on the self-attention layers, each encoder attends to the corresponding modality to compute the attention scores. In this model, we have used cross-modal attention layers, which also compute the dot product between the K and Q matrices, but these are taken from different modalities as shown in Fig. 9.3. By introducing cross-modal attention layers, we are able to capture the complementary relationships between the audio and visual modalities. After obtaining the attended features, the audio and visual features are concatenated to obtain the segment-level A-V feature representations. Finally, the ASP mechanism (Okabe et al. 2018) is employed to compute the weighted first- and second-order statistics (i.e., mean μ and standard deviation σ) which are then concatenated to attain the final A-V embedding.

9.4 Experiments

9.4.1 Dataset and Evaluation Metric

We have evaluated the proposed model on the Voxceleb1 dataset (Nagrani et al. 2020), obtained from YouTube videos under challenging environments. The dataset consists of 148,642 video clips, captured from 1,251 speakers, of which 55% are male. Each video clip has a duration of 4 to 145 seconds and the speakers are chosen to cover a diverse range of ethnicities, accents, professions, and ages. For our experiments, we divided the voxceleb1 development set, which has 1,211 speakers into training and validation sets. The training and validation splits were randomly selected as 1,150 and 61 speakers, respectively, and we report our results on both the validation split and Vox1-O (Voxceleb1 original) test set. It is worth mentioning that the model is trained only on the training set of the voxceleb1 dataset.

Equal Error Rate (EER) is used as an evaluation metric, which refers to the point where the False Accept Rate (FAR) is equal to the False Reject Rate (FRR). A perfect scoring model should yield an EER of zero, so a lower EER value indicates better performance.

9.4.2 Experimental Setup

In the case of the audio modality, the ECAPA-TDNN (Desplanques et al. 2020) framework is used for training the audio-based speaker embedding network. The 80-dimensional Mel-FilterBank (MFB) features, which are extracted using an analysis window size of 25 ms over a frame-shift of 10 ms, are used as input acoustic features to the network. The audio data are randomly augmented on-the-fly with either MUSAN noise, speed perturbation with a rate between 0.95 and 1.05, or reverberation (Snyder et al. 2015). In addition, we use SpecAugment (Ko et al. 2017) to apply frequency and time masking on the MFB features. The weights of the embedding network are initialized with values from the normal distribution and the network is trained for a maximum of 100 epochs, and early stopping is used. The network is optimized using the Adam optimizer (Kingma and Ba 2015) with the initial learning rate of 0.001 and the batch size fixed to 400. In order to prevent the network from overfitting, dropout is used with p = 0.5 after the last linear layer. Finally, a weight decay of $5e - 4$ is used for all experiments.

For the visual modality, the face embedding network is trained using the Resnet-18 architecture. For regularizing the network, dropout is used with $p = 0.8$ on the linear layers. The initial learning rate of the network which was set to $1e - 2$ is used for the Adam optimizer. Weight decay of $5e - 4$ is used. The batch size of the network is set to 400. Data augmentation is performed on the training data by random cropping, which produces a scale-invariant model. The number of epochs is set to 50 and early stopping is used to obtain the best weights of the network.

As for the fusion network, we used hyperbolic tangent functions for the activation of cross-attention modules. The dimension of the extracted features of the audio modality is set to 192, and 512 for the visual modality. In the joint cross-attention module, the initial weights of the joint cross-attention matrix are initialized with the Xavier method (Glorot and Bengio 2010) and the weights are updated using the Adam optimizer. The initial learning rate is set to 0.001 and batch size is fixed to 100. A dropout of 0.5 is applied on the attended A-V features, and a weight decay of $5e - 4$ is used for all experiments.

Table 9.1 SV performance of the RJCA fusion approach on the validation set with varied number of recursions t. Results are reported in terms of EER

Fusion method	Validation set
RJCA Fusion ($t = 2$)	2.029
RJCA Fusion ($t = 3$)	**1.946**
RJCA Fusion ($t = 4$)	1.995
RJCA Fusion ($t = 5$)	2.159

9.4.3 *Results*

In order to evaluate the performances of different variants of CA-based A-V SV systems, a series of experiments was conducted on the VoxCeleb1 dataset. Results are reported on the validation and VoxCeleb1 Original (Vox1-O) test sets in terms of EER based on the average of three runs for statistical stability. To obtain the audio and visual feature vectors, we used Resnet-18 (He et al. 2016) for the visual modality and ECAPA-TDNN (Desplanques et al. 2020) for audio, similar to Rajasekhar and Alam (2023) and Tao et al. (2023) in order to have a fair comparison.

9.4.3.1 Ablation Study with Varying Number of Recursions for RJCA

To analyze the contribution of the individual components of the Recursive Joint Cross-Attention (RJCA)-based fusion approach, we carried out a set of experiments by varying the number of iterations of the recursive mechanism. The results of this ablation study is presented in Table 9.1. We can observe from these results that the best performance is achieved at 3 iterations. Beyond that, we observe a decline in the fusion performance, which can be attributed to the fact that the recursion acts as a regularizer and improves the generalization ability of the RJCA model. Note that the results are reported based on the average of three runs for statistical stability.

9.4.3.2 Comparative Evaluation of CA-Based mechanisms

In order to understand the potential of CA-based mechanisms, we compared multiple variants of CA-based mechanisms and provide insight on the impact of effectively capturing the intra- and inter-modal relationships across audio and visual modalities, respectively. We also provided a comparison of some of the widely used fusion strategies for a comprehensive analysis of audio-visual fusion for speaker verification. Table 9.2 presents the results of CA-based fusion mechanisms along with some of the widely-used fusion strategies on the validation set in terms of EER. We first implemented a simple score-level fusion, where scores obtained from individual modalities are fused. Next, we explored feature concatenation, where the deep features of audio and visual modalities are concatenated and the joint A-V representation is then used to obtain the final verification score. From the reported results

Table 9.2 A comparison of A-V speaker verification performance of the introduced fusion approaches with some existing fusion strategies. The results are reported in terms of EER on the validation set

Fusion method	Validation set
Score-level Fusion	2.521
Feature Concatenation	2.489
Self-Attention (SA)	2.412
Cross-Attention (CA)	2.387
Joint Cross-Attention (JCA)	2.125
RJCA	**1.946**
CMT	**1.593**

in Table 9.2, we can observe that the fusion performance improves over simple score-level fusion. By employing a self-attention mechanism on the concatenated features of audio and visual modalities, the fusion performance improves further, due to the fact the self-attention mechanism helped to leverage the intra-modal relationships. We also explored inter-modal relationships across the modalities using CA and achieved further improvement in fusion performance as provided in Table 9.2.

Then we explored a Joint Cross-Attention (JCA) mechanism, where a joint A-V feature representation is deployed in the CA framework to simultaneously capture both intra- and inter-modal relationships between audio and visual modalities. Again, we explored an extended version of JCA called "Recursive JCA" (RJCA), where we introduced the recursive fusion of the attended features of individual modalities and observed that recursive fusion with 3 iterations to obtain more refined feature representations and thus better performance than with JCA, CA, SA, feature concatenation and score-level multimodal fusion strategies. Finally, we performed experiments with the proposed Cross-Modal Transformer fusion technique. As we can observe in Table 9.2, using a joint feature representation significantly improves the performance over unimodal approaches as they can simultaneously capture intra- and inter-modal relationships across the audio and visual modalities. Multimodal Transformers based on cross-attention further improved performance.

Table 9.3 Performance of the newly introduced fusion approaches in comparison to existing state-of-the-art fusion strategies. EER is used for reporting the results on the validation and test sets of the VoxCeleb1 corpus

Fusion method	Validation set	Vox1-O set
Visual	3.720	3.779
Audio	2.553	2.529
Tao et al. (2023)	2.476	2.409
JCA	2.125	2.214
RJCA	1.946	**2.015**
CMT	1.851	**1.716**

9.4.3.3 Comparison to the State-of-the-Art

Most of the existing methods used the Voxceleb2 development dataset for training SV models. However, in this work, we used the Voxceleb1 dataset to validate the proposed approach and compared it against recent state-of-the-art SVmethods using A-V fusion. Table 9.3 shows the comparison of performance of our introduced fusion approaches (namely RJCA and CMT) to recent state-of-the-art methods as well as individual modalities on both validation split of Voxceleb1 and Vox1-O test sets. First, we conducted experiments with the individual modalities and observed that the audio-only SV system performed relatively better than the one based on the visual modality. In order to ensure a fair comparison, we re-implemented the approach of Rajasekhar and Alam (2023) and Tao et al. (2023) using the same experimental setup on the Voxceleb1 dataset. Tao et al. (2023) explored the complementary relationships as supervisory information to deal with noisy samples. The JCA approach initially proposed in Rajasekhar and Alam (2023) deployed a joint A-V representation in the CA framework and improved fusion performance. Since the RJCA approach employs recursive fusion with the joint cross-attentional framework to obtain robust feature representations, we can observe from Table 9.3 that fusion performance improves further. It is evident from Table 9.3 that the JCA, RJCA and Cross-Modal Transformer (CMT)-based fusion models outperform all the considered baseline systems both on the validation and test sets. The proposed CMT fusion approach achieved the best performance, providing a relative improvement of approximately 54%, 32%, 28%, 22% and 14% over the Visual-only, Audio-only, Tao et al. (2023), JCA, and RJCA fusion approaches, respectively, on the Vox1-O test set of the VoxCeleb1 dataset.

9.5 Conclusion

In this research, we investigated three variants of cross-attention-based multimodal fusion models, namely Joint Cross-Attention (JCA), Recursive JCA (RJCA) and Cross-Modal Transformer (CMT), for A-V SV. In particular, we obtained the deep features of audio and visual modalities from pre-trained networks, which were then fed to the fusion model. After that, the semantic relationships between audio and visual modalities were obtained based on the cross-correlation between the individual feature representations. Unlike JCA, the RJCA approach effectively exploited both inter- and intra-modal relationships across audio and visual modalities in an iterative manner in order to obtain more refined A-V feature representations. The CMT fusion model was able to effectively leverage the complementary inter-modal relationships between the audio and visual modalities. Experimental A-V speaker verification results on the VoxCeleb1 dataset demonstrated the effectiveness of the JCA, RJCA and CMT multimodal fusion strategies. The performance of these approaches can be further enhanced by training with the large-scale Voxceleb2 dataset as it can improve the generalization capability still further.

Acknowledgements The authors wish to acknowledge the funding from the Government of Canada's New Frontiers in Research Fund (NFRF) through grant NFRFR-2021-00338. Any opinions, findings, conclusions or recommendations expressed in this material are those of the authors and do not necessarily reflect those of the NFRF.

References

Alam J (2023) On the use of cross- and self-module attentive statistics pooling techniques for text-independent speaker verification. In: 2023 IEEE international joint conference on biometrics (IJCB), pp 1–9

Alam J, Boulianne G, Burget L, Dahmane M, Diez MS, Glembek O, Lalonde M, Lozano AD, Matějka P, Mizera P, Mošner L, Noiseux C, Monteiro J, Novotný O, Plchot O, Rohdin AJ, Silnova A, Slavíček J, Stafylakis T, St-Charles PL, Wang S, Zeinali H (2020) Analysis of abc submission to nist sre 2019 cmn and vast challenge. In: Proceedings of Odyssey 2020 the speaker and language recognition workshop, international speech communication association, vol 11, pp 289–295. https://www.fit.vut.cz/research/publication/12292

Chen H, Zhang H, Wang L, Lee KA, Liu M, Dang J (2023) Self-supervised audio-visual speaker representation with co-meta learning. In: International conference on acoustics, speech, and signal processing (ICASSP), pp 1–5

Chen T, Kornblith S, Norouzi M, Hinton G (2020) A simple framework for contrastive learning of visual representations. In: Proceedings of the 37th international conference on machine learning, proceedings of machine learning research, vol 119, pp 1597–1607. https://proceedings.mlr.press/v119/chen20j.html

Deng J, Guo J, Zafeiriou S (2018) Arcface: additive angular margin loss for deep face recognition. In: 2019 IEEE/CVF conference on computer vision and pattern recognition (CVPR), pp 4685–4694. https://api.semanticscholar.org/CorpusID:8923541

Desplanques B, Thienpondt J, Demuynck K (2020) ECAPA-TDNN: emphasized channel attention. Propagation and aggregation in TDNN based speaker verification. In, The international speech communication association (INTERSPEECH), pp 3830–3834

Duan B, Tang H, Wang W, Zong Z, Yang G, Yan Y (2021) Audio-visual event localization via recursive fusion by joint co-attention. In: IEEE WACV, pp 4012–4021

Glorot X, Bengio Y (2010) Understanding the difficulty of training deep feedforward neural networks. ACAIS 9:249–256

Guo MH, Xu TX, Liu JJ, Liu ZN, Jiang PT, Mu TJ, Zhang SH, Martin RR, Cheng MM, Hu SM (2022) Attention mechanisms in computer vision: a survey. Comput Vis Media 3:331–368. http://dx.doi.org/10.1007/s41095-022-0271-y

Hansen JH, Hasan T (2015) Speaker recognition by machines and humans: a tutorial review. IEEE Signal Process Mag 32(6):74–99

He K, Zhang X, Ren S, Sun J (2016) Deep residual learning for image recognition. In: Conference on computer vision and pattern recognition (CVPR), pp 770–778

Hörmann S, Moiz A, Knoche M, Rigoll G (2020) Attention fusion for audio-visual person verification using multi-scale features. In: 2020 15th IEEE international conference on automatic face and gesture recognition (FG 2020), pp 281–285. https://api.semanticscholar.org/CorpusID:231683117

Jiang MX, Ji S (2022) Cross-modality gated attention fusion for multimodal sentiment analysis. arXiv:abs/2208.11893. https://api.semanticscholar.org/CorpusID:251800208

Kingma DP, Ba J (2015) Adam: a method for stochastic optimization. In: 3rd international conference on learning representations, ICLR 2015, San Diego, CA, USA, May 7–9, 2015, Conference track proceedings. http://arxiv.org/abs/1412.6980

Ko T, Peddinti V, Povey D, Seltzer ML, Khudanpur S (2017) A study on data augmentation of reverberant speech for robust speech recognition. In: International conference on acoustics, speech, and signal processing (ICASSP), pp 5220–5224

Kretov AP (2020) Attention mechanism in natural language processing. PhD thesis, Czech Technical University in Prague. https://api.semanticscholar.org/CorpusID:219964901

Lee JT, Jain M, Park H, Yun S (2021) Cross-attentional audio-visual fusion for weakly-supervised action localization. In: Proceedings of the ICLR

Liu M, Lee KA, Wang L, Zhang H, Zeng C, Dang J (2023) Cross-modal audio-visual co-learning for text-independent speaker verification. In: International conference on acoustics, speech, and signal processing (ICASSP), pp 1–5

Mocanu BC, Tapu R (2022) Active speaker recognition using cross attention audio-video fusion. In: 2022 10th European workshop on visual information processing (EUVIP), pp 1–6. https://api.semanticscholar.org/CorpusID:253047036

Nagrani A, Albanie S, Zisserman A (2018a) Learnable PINs: cross-Modal Embeddings for Person Identity. In: European conference on computer vision. https://api.semanticscholar.org/CorpusID:13746939

Nagrani A, Albanie S, Zisserman A (2018b) Seeing voices and hearing faces: cross-modal biometric matching. In: 2018 IEEE/CVF conference on computer vision and pattern recognition, pp 8427–8436. https://api.semanticscholar.org/CorpusID:4559198

Nagrani A, Chung JS, Xie W, Zisserman A (2020) Voxceleb: large-scale speaker verification in the wild. Comput Speech Lang 60:101027. https://www.sciencedirect.com/science/article/pii/S0885230819302712

Okabe K, Koshinaka T, Shinoda K (2018) Attentive statistics pooling for deep speaker embedding. In: The international speech communication association (INTERSPEECH), pp 2252–2256

Praveen RG, Alam J (2024a) Audio-visual person verification based on recursive fusion of joint cross-attention. IEEE conference on automatic face and gesture recognition (IEEE FG). arXiv: abs/2403.04654. https://api.semanticscholar.org/CorpusID:268264097

Praveen RG, Alam J (2024b) Cross-modal transformers for audio-visual person verification. In: Proceedings of the speaker and language recognition workshop (Odyssey)

Praveen RG, Granger E, Cardinal P (2021) Cross attentional audio-visual fusion for dimensional emotion recognition. In: 2021 16th IEEE international conference on automatic face and gesture recognition (FG 2021), pp 1–8. https://api.semanticscholar.org/CorpusID:243860821

Praveen RG, de Melo WC, Ullah N, Aslam H, Zeeshan O, Denorme T, Pedersoli M, Koerich AL, Bacon S, Cardinal P, Granger E (2022) A joint cross-attention model for audio-visual fusion in dimensional emotion recognition. In: Proceedings of the IEEE/CVF conference on computer vision and pattern recognition (CVPR) workshops, pp 2486–2495

Praveen RG, Cardinal P, Granger E (2023) Audio-visual fusion for emotion recognition in the valence-arousal space using joint cross-attention. IEEE Trans Biomet, Behav, Ident Sci 5(3):360–373

Praveen RG, Granger E, Cardinal P (2023b) Recursive joint attention for audio-visual fusion in regression based emotion recognition. In: International conference on acoustics, speech, and signal processing (ICASSP), pp 1–5

Qian Y, Chen Z, Wang S (2021) Audio-visual deep neural network for robust person verification. IEEE/ACM Trans Audio, Speech, Lang Process 29:1079–1092

Rajasekhar GP, Alam J (2023) Audio-visual speaker verification via joint cross-attention. Speech Comput. Springer Nature Switzerland, Cham, pp 18–31

Sarı L, Singh K, Zhou J, Torresani L, Singhal N, Saraf Y (2021) A multi-view approach to audio-visual speaker verification. In: International conference on acoustics, speech, and signal processing (ICASSP), pp 6194–6198

Seyed, Greenberg C, Singer E, Olson D, Mason L, Hernandez-Cordero J (2020) The 2019 nist audio-visual speaker recognition evaluation. In: The speaker and language recognition workshop: Odyssey 2020, Tokyo. https://tsapps.nist.gov/publication/get_pdf.cfm?pub_id=929541

Shon S, Oh TH, Glass J (2019) Noise-tolerant audio-visual online person verification using an attention-based neural network fusion. In: International conference on acoustics, speech, and signal processing (ICASSP), pp 3995–3999

Snyder D, Chen G, Povey D (2015) MUSAN: A music, speech, and noise corpus. CoRR arXiv:abs/1510.08484. http://arxiv.org/abs/1510.08484

Sun P, Zhang S, Liu Z, Yuan Y, Zhang T, Zhang H, Hu P (2023) A method of audio-visual person verification by mining connections between time series. Proc Interspeech 2023:3227–3231

Tao R, Lee KA, Shi Z, Li H (2023) Speaker recognition with two-step multi-modal deep cleansing. In: International conference on acoustics, speech, and signal processing (ICASSP), pp 1–5

Tu Y, Lin W, Mak MW (2022) A survey on text-dependent and text-independent speaker verification. IEEE Access 10:99038–99049

Wang M, Deng W (2021) Deep face recognition: a survey. Neurocomputing 429:215–244

Wang W, Tran D, Feiszli M (2020) What makes training multi-modal classification networks hard? In: Conference on computer vision and pattern recognition (CVPR), pp 12692–12702

Zhang H, Wu W (2022) Transformer gate attention model: An improved attention model for visual question answering. In: 2022 international joint conference on neural networks (IJCNN), pp 1–7

Zhao D, Chang Z, Guo S (2019) A multimodal fusion approach for image captioning. Neurocomputing 329:476–485. https://www.sciencedirect.com/science/article/pii/S0925231218313213

Chapter 10
An Efficient Clustering Algorithm for Self-Supervised Speaker Recognition

Abderrahim Fathan and Jahangir Alam

Abstract Clustering-based pseudo-labels (PLs) are widely used to optimize speaker embedding (SE) networks and train self-supervised speaker verification (SV) systems. However, PL-based self-supervised training depends on high-quality PLs, and clustering performance relies heavily on time- and resource-consuming data augmentation regularization. In this chapter, we propose an efficient and general-purpose multi-objective clustering algorithm that outperforms all other baselines for clustering SEs. Our approach, called Contrastive Information Maximization Clustering (CIMC), avoids explicit data augmentation, enabling fast training with low memory and compute resource usage. It is based on three principles: (1) self-augmented training to enforce representation invariance and maximize the information-theoretic dependency between samples and their predicted PLs; (2) virtual mixup training to impose local-Lipschitzness and enforce the cluster assumption; and (3) supervised contrastive learning to learn more discriminative features, pulling samples of the same class together and pushing samples of different classes apart, while improving robustness to natural corruptions. We provide a thorough comparative analysis of the performance of our clustering method against baselines using a variety of clustering metrics, demonstrating that we outperform all other clustering benchmarks. Moreover, we perform an ablation study to analyze the contribution of each component, including two other augmentation-based objectives, and show that our multi-objective approach provides beneficial complementary information. Finally, using the generated PLs to train our SE system enables us to achieve SOTA SV performance.

10.1 Introduction

Speaker Verification (SV) is the task of confirming the identity of a speaker based on the speaker's known utterances (for details refer to Sect. 9.1). In recent years, it has become a key technology for personnel authentication in numerous applications

A. Fathan (✉) · J. Alam
Computer Research Institute of Montreal, Montreal (Quebec), Canada
e-mail: abderrahim.fathan@crim.ca

© The Author(s), under exclusive license to Springer Nature Switzerland AG 2025

P. Passban et al. (eds.), *Enhancing LLM Performance*, Machine Translation: Technologies and Applications 7, https://doi.org/10.1007/978-3-031-85747-8_10

(Hansen and Hasan 2015). Typically, utterance-level fixed-dimensional embedding vectors are extracted from the enrollment and test speech samples and then fed into a scoring algorithm (e.g., cosine distance) to measure their likelihood of being from the same speaker. Classically, the i-vector framework has been one of the most dominant approaches for speech embedding (Dehak et al. 2011; Kenny 2012) thanks to its ability to summarize the distributive patterns of speech in an unsupervised manner and with relatively small training datasets. It generates fixed-sized compact vectors that represent the speaker's identity in a speech utterance regardless of its length. In addition, various deep learning-based architectures and techniques have been proposed to extract embeddings (Bai and Zhang 2021; Fathan et al. 2022; Kang et al. 2022). They have shown great performance when large training datasets are available, particularly with a sufficient number of speakers (Snyder et al. 2018). One widely employed architecture for this purpose is ECAPA-TDNN (Desplanques et al. 2020), which has achieved state-of-the-art (SOTA) performance in text-independent speaker recognition. The latter uses Squeeze-and-Excitation (SE), employs channel- and context-dependent statistics pooling and multi-layer aggregation and applies self-attention pooling to obtain an utterance-level embedding vector.

Indeed, most of the deep embedding models are trained in a fully supervised manner and require large speaker-labeled datasets for training. However, well-annotated datasets can be expensive and time-consuming to prepare, which has led the research community to explore more affordable Self-Supervised Learning (SSL) techniques using larger unlabeled datasets. One common way to solve this issue for SV systems is to use a one-stage "clustering-classification" scheme (Fathan et al. 2022; Kang et al. 2022) by employing clustering algorithms (e.g., K-means, agglomerative hierarchical clustering, spectral clustering) or other SSL-based objectives (e.g., SimCLR or MoCo (Xia et al. 2021)) to generate pseudo-labels (PLs) and train the speaker embedding network using these labels in a discriminative fashion. More recently, better-performing ways have emerged which are now widely adopted in the SV domain. These frameworks

Despite the impressive performance of these PL-based self-supervised SV schemes, clustering performance remains a bottleneck in all above approaches (Han et al. 2022; Tao et al. 2022). Indeed, downstream performance relies greatly on accurate PLs since these are generally noisy and inaccurate due to the discrepancy between the clustering objectives and the final SV task. Besides, even with iterative clustering-classification paradigms, erroneous information from the wrong PLs keeps propagating iteratively, which degrades the final performance (Li et al. 2021; Tao et al. 2022). Thus, there is a need for better-performing clustering algorithms to generate less noisy and more accurate PLs. Rather than using SOTA deep clustering models which rely on heavy domain-specific data augmentations, these approaches usually employ classical clustering algorithms such as K-means or spectral clustering because these are easier to use, faster, and less resource-consuming in terms of memory and GPU/CPU resources to train.

We propose Contrastive Information Maximization Clustering (CIMC), an efficient and general-purpose multi-objective clustering algorithm that outperforms all other baselines used for clustering speaker embeddings. Our approach avoids using explicit data augmentation for fast training and low memory and compute resource usage. Our solution is based on the combination of three principles: (1) self-augmented training to enforce representation invariance and maximize the information-theoretic dependency between samples and their predicted PLs through the Information Maximizing Self-Augmented Training (IMSAT) clustering framework (Hu et al. 2017); (2) virtual mixup training (VMT) (Mao et al. 2019) to impose local-Lipschitzness which enforces the cluster assumption; and (3) supervised contrastive learning (Khosla et al. 2020) by leveraging on-the-fly generated PLs, to pull samples of the same class together and push samples of different clusters apart.

Instead of mixing up inputs or using contrastive loss solely to enforce smoother model responses and compactness of the embeddings, our CIMC method leverages these predictions as additional supervisory signals to better guide cluster assignment for more robust, stable, and better-performing data clustering.

The contributions of this chapter are as follows:

- We propose a novel general-purpose multi-objective clustering algorithm, called CIMC, for large-scale datasets or/and a high number of clusters.
- We explore various recent SOTA SSL objectives for clustering where we show that multi-objective clustering often provides beneficial complementary information.
- Our proposed method outperforms a large set of clustering baselines. Additionally, by using the generated labels to train our SV systems, we achieve high SV performance. Furthermore, employing augmentation-based SSL objectives allowed us to achieve both SOTA speaker embedding clustering and SV performance.

10.2 Background and Related Work

Diverse methods for clustering have been proposed such as K-means (Hartigan and Wong 1979), Gaussian mixture model (GMM), BIRCH (Zhang et al. 1997), CURE (Guha et al. 1998), and Agglomerative Hierarchical Clustering (AHC) (Day et al. 1984). However, these methods can only fit linear boundaries between data representations. Recently, the powerful representative ability of deep neural networks (DNNs) has been leveraged to model the non-linearity of complex data and to scale to large datasets. For instance, Deep Embedded Clustering (DEC) (Xie et al. 2016) proposed to use deep models to simultaneously learn feature representations and cluster assignments, while the DeepCWRN (Dahal 2018) approach employs an autoencoder to simultaneously learn feature representations and embeddings suitable for clustering by encouraging the separation of natural clusters in the embedding space. Beside these, other deep models have been proposed based on generative models (Dilokthanakul et al. 2016; Jiang et al. 2016) or dynamic architectures (Ronen et al. 2022).

While data augmentation remains a crucial component to regularize DNNs for clustering and unsupervised representation learning in order to model the invariance of learned representations (Dosovitskiy et al. 2014), augmentation has the downside of increasing the training set which can lead to significantly more training time. Especially for large-scale datasets and neural networks. Additionally, using blind augmentations can negatively affect speaker verification/recognition tasks because transformations like pitch perturbation or spectral augmentation can alter the identity of a speaker, potentially creating misleading data samples. Moreover, for real-world tabular data applications (Bahri et al. 2021) such as genomics and clinical data, generating additional augmented views is not an obvious task and can be prohibited.

Finally, while our work focuses on generating highly accurate annotations for audio-based self-supervised SV, we believe our clustering approach and insights from studying several SSL-based objectives can be useful for multi-modal approaches that combine information from different available views/sources in a self-supervised manner, such as audio and visual modalities. In fact, our proposed clustering objective can be adapted to other types of input data and used to leverage/fuse complementary annotations to train SSL-based models. Furthermore, we find our explored methods and concepts could potentially be applied to the development of multi-modal LLMs.

10.3 Proposed Clustering Approach

A schematic diagram of the proposed CIMC is presented in Fig. 10.1, which is trained via minimizing the total loss $\mathcal{L}_{total}$, that integrates three different SSL-based loss functions. The aim is to harness these objectives as additional supervisory signals to regularize the clustering model to produce consistent assignments.

Given a DNN-based clustering model f and a predefined number of clusters C, the CIMC approach constrains the predictions of the model to remain unchanged under local perturbations and implicit Virtual Mixup Training (VMT) data augmentations ($\mathcal{L}Mixup$) (Mao et al. 2019). The model is trained end-to-end by imposing local-Lipschitzness on the learned weights to favor the cluster assumption (if samples are in the same cluster, they come from the same class), which is a critical condition for successful clustering (Grandvalet and Bengio 2004). More explicitly, it optimizes the following $\mathcal{L}_{total}$ objective, as shown in Eq. (10.1):

$$\mathcal{L}_{total} = \mathcal{L}_{IMSAT} + \mathcal{L}_{SupCon} + \mathcal{L}_{Mixup} \tag{10.1}$$

where $\mathcal{L}_{IMSAT}$ is the loss function for the original IMSAT clustering objective and $\mathcal{L}_{SupCon}$ and $\mathcal{L}_{Mixup}$ represent the supervised contrastive loss and the mixup loss terms, respectively. Our main focus is to harness these objectives as additional supervisory signals to regularize the clustering model to produce consistent assignments.

The original IMSAT loss $\mathcal{L}_{IMSAT}$ (Hu et al. 2017) is expressed mathematically as in Eq. (10.2):

$$\mathcal{L}_{IMSAT} = R_{SAT}(\theta, T_{VAT}) + \lambda(H(Y|X) - \mu H(Y)), \tag{10.2}$$

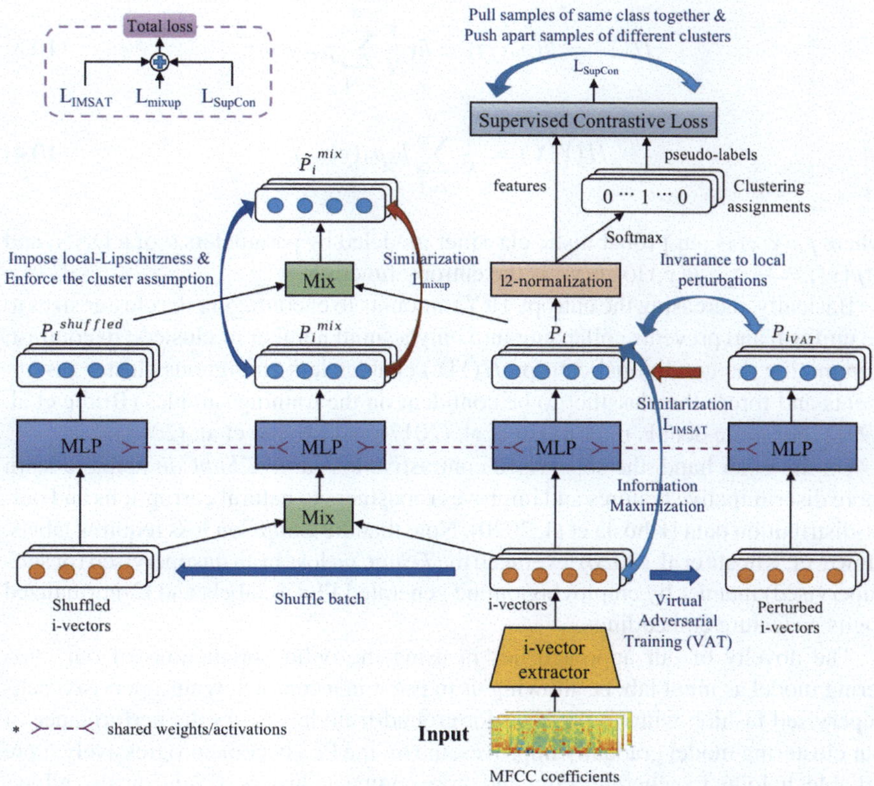

Fig. 10.1 The pipeline of our proposed CIMC method, depicting the data flow and the different losses employed for clustering

where $H(.)$ and $H(.|.)$ are the marginal and conditional entropy, respectively, and the loss term $R_{SAT}(\theta; T) = \frac{1}{N}\sum_{n=1}^{N} R_{SAT}(\theta; x_n, T(x_n))$ allows the representations of the augmented samples to be further pushed close to those of the original samples while also regularizing the complexity of the network against local perturbations using Virtual Adversarial Training (VAT) (Miyato et al. 2018). $\lambda, \mu \in \mathbb{R}$ are hyperparameters that control the trade-offs between the complexity regularization of the model (through R_{SAT}) and the Mutual Information (MI) maximization, and between the two entropy terms, respectively. $R_{SAT}(\theta; x, T(x)) = -\sum_{c=1}^{C}\sum_{y_c=0}^{1} p_{\hat{\theta}}(y_c|x) log p_{\theta}(y_c|T(x))$, where $p_{\hat{\theta}}(y_c|x)$ is the prediction of original data point x, and $\hat{\theta}$ are the current parameters of the network. $T_{VAT}(x) = x + r$ is the augmentation function using local perturbations to enforce invariance where $r = \arg\max_{r'}\{R_{SAT}(\hat{\theta}; x, x + r'); \|r'\|_2 \leq \epsilon\}$ is an adversarial direction.

The difference between the marginal and conditional entropy represents the MI between sample X and label Y, which we maximize. The two entropy terms can be calculated as in Eqs. (10.3) and (10.4):

$$H(Y) = h(p_\theta(y)) = h(\frac{1}{N}\sum_{i=1}^{N} p_\theta(y|x)), \tag{10.3}$$

$$H(Y|X) = \frac{1}{N}\sum_{i=1}^{N} h(p_\theta(y|x_i)), \tag{10.4}$$

where $p_\theta(y|x)$ is our probabilistic classifier modeled by parameters θ of a DNN, and $h(p(y)) = -\sum_{y'} p(y') \log p(y')$ is the entropy function.

Basically, increasing the entropy H(Y) amounts to encouraging the cluster sizes to be uniform and prevents collapsing into only a small number of clusters; in contrast, minimizing the conditional entropy $H(Y|X)$ enables less ambiguous cluster assignments and forces the classifier to be confident on the training samples (Bridle et al. 1991). For more details refer to Hu et al. (2017) and Miyato et al. (2018).

On the other hand, the supervised contrastive loss term $\mathcal{L}SupCon$ helps to learn more discriminative features and improves robustness to natural corruptions and out-of-distribution data (Khosla et al. 2020). Note that the $\mathcal{L}SupCon$ loss requires labels. However, Khosla et al. (2020) leveraged the $\mathcal{L}SupCon$ loss in an unsupervised (or self-supervised) manner by employing online generated PLs as labels and l2-normalized logits as feature embeddings.

The novelty of our approach lies in using the online predictions of our clustering model as input labels, allowing us to use it in a completely unsupervised/self-supervised fashion without the need for ground-truth labels. As the performance of our clustering model gradually improves, the online PLs become progressively more reliable, helping to generate better and more compact clusters. $\mathcal{L}SupCon$ also allows us to leverage online clustering assignments by using nearest-neighbors as positives rather than augmentations, pulling clusters of points belonging to the same class together in embedding (logit) space while simultaneously pushing apart clusters of samples from different classes. Further mathematical details and discussion about the $\mathcal{L}_{SupCon}$ objective are included in Sect. 10.7.

The mixup loss term $\mathcal{L}_{Mixup}$ can be formulated as in Eq. (10.5):

$$\mathcal{L}_{Mixup} = \frac{1}{N}\sum_{i=1}^{N} KL(\alpha_i p_i + (1-\alpha_i)p_{r_i} || f(\alpha_i x_i + (1-\alpha_i)x_{r_i})). \tag{10.5}$$

where N is the size of data (or mini-batches), $r_i \in \{1, .., N\}$ is a random index, and $\alpha_i \in [0, 1]$ is the mixup interpolation coefficient. $KL(.||.)$ refers to the Kullback-Leibler divergence. $p_i = f(x_i) \in \mathbb{R}^{1xC}$, $p_{r_i} = f(x_{r_i})$ correspond to the predictions of data samples x_i and x_{r_i}. Finally, inspired by the VMT regularization method (Mao et al. (2019)), which encourages the model to behave linearly between training points, we enforce representation smoothness during clustering and enforce consistent predictions between the surrounding and training points. Indeed, mixup (Zhang et al. 2017) which is a strategy to augment data by interpolating different data samples alongside their labels, often leads to better generalization to out-of-set samples.

Mixup was also found by Fathan et al. (2022) to lead to better generalization of self-supervised SV systems when the clusters are not compact or well-distanced, as it can dilute label noise and induce better class separation. Following the work of Mao et al. (2019), instead of directly mixing probabilities in the $\mathcal{L}_{Mixup}$ loss, we perform mixup over logits, followed by Softmax for better training and to prevent early information loss during the mix of probabilities. During experiments, we found this to considerably improve results and convergence compared to mixup on probabilities.[1]

10.4 Discussion of the Multi-Objective Clustering Approach

Our approach in this chapter aims to extend and improve the IMSAT method by incorporating and studying additional SSL objectives in a multi-objective fashion within the current framework. Our study investigates useful self-supervised objective losses for the purpose of speaker clustering and recognition. To this aim, we harness these objectives as additional supervisory signals during clustering to regularize the clustering model to produce consistent feature representations. Additionally, this combination of objectives can increase the model's expressiveness through various inductive biases, maximize the amount of information learned per sample, and help it learn weights that can better disambiguate difficult or complex data examples. This can also enable our clustering model to self-correct its early mislabelling, and reduce the likelihood of learning spurious features since in that case the weights are constrained to simultaneously satisfy all the training objectives (i.e., assume the simplest hypothesis). Our work also analyzes the complementary information between these objectives using our large MLP-based architecture, without interference from other architectural biases.

The IMSAT framework, which is the backbone of our clustering approach helps to avoid degenerate solutions that other clustering methods are susceptible to by being rigorously grounded in information theory. Indeed, due to the entropy maximisation component $H(Y)$ within MI, the loss objective is not minimised if all inputs are assigned to the same class. At the same time, to minimize the MI loss term, it is optimal for the model to predict for each input a single class with certainty (i.e., one-hot) due to the additional conditional entropy component $H(Y|X)$ that we minimize. Hence, we avoid clusters disappearing during training or a single cluster starts dominating the predictions. During our experiments, we find the $\mathcal{L}_{IMSAT}$ loss to be critical for good clustering performance.

Inspired by the VMT regularization method (Mao et al. 2019), which encourages the model to behave linearly between training points, we enforce representation smoothness during clustering and enforce consistent predictions between the surrounding and training points (Verma et al. 2022). The mixup method (Zhang et al. 2017), as an efficient strategy to augment data by interpolating different data samples

[1] Code of our clustering framework is available at: https://github.com/fathana/CAMSAT_clustering

alongside their labels, often leads to better generalization to out-of-set samples. It has proven its strength in various tasks (e.g., image classification (Zhang et al. 2017), anti-spoofing (Tomilov et al. 2021) and speech recognition (Meng et al. 2021)). (Zhang et al. 2017) has shown that mixup not only reduces the memorization to adversarial samples, but also performs better than Empirical Risk Minimization (Vapnik et al. 2015). Mixup has also been found by Fathan et al. (2022) to be effective against label noise memorization (Arpit et al. 2017) and to lead to better generalization of self-supervised SV systems when the clusters are not compact or well-distanced. We find this property to be especially important during the clustering of speaker embeddings to mitigate strong label noise in the early training epochs and avoid early convergence to suboptimal cluster assignments. As mixup can dilute label noise in online generated PLs and create synthetic samples around the borders—leading to smoothing the data manifold and better class separation—we believe this can help slow down the memorization of noisy PLs and allow the model to learn from the simple patterns available for a longer period. Consequently, this leads to better clusters and induces robustness, improved generalization capability, and better online clustering stability for large-scale datasets.

The resulting CIMC algorithm is highly scalable, fast, more robust than IMSAT to corruptions and shifts in the data during online clustering, is simple to implement, yet adds only limited computational overhead to IMSAT. Additionally, we believe CIMC can be significantly beneficial in further optimizing current self-supervised SV frameworks by replacing simple clustering methods (e.g., k-means, spectral clustering). It can also be used in speaker diarization frameworks to improve the clustering aspects of speaker diarization methods where clustering is one of the important modules. Finally, our proposed clustering approach is a general-purpose method and can be applied to problems and domains other than speech.

10.5 System Description

10.5.1 Clustering-Based Self-Supervised Speaker Embedding Framework

Figure 10.2 shows a schematic diagram of our general clustering-based self-supervised SV process that we follow throughout the chapter. In this work, we explore various clustering algorithms and analyze the impact of pretraining, particularly clustering, on SV performance. We employ ECAPA-TDNN as our speaker embedding network and use AAMSoftmax (Deng et al. 2018) objective loss to train our systems using PLs generated by various clustering algorithms.

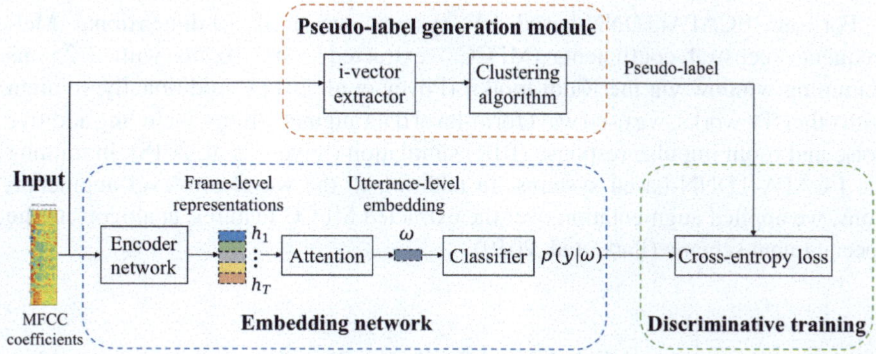

Fig. 10.2 General process for training our clustering generated pseudo-label-based self-supervised speaker embedding networks

10.5.2 Clustering-Based Pseudo-Label Generation

For clustering, we extracted i-vectors (Dehak et al. 2011; Kenny 2012) using the Kaldi toolkit (Povey et al. 2011), which provides a statistical, unsupervised, fixed-dimensional representation from each training utterance. Clustering was then performed on these i-vectors. After training the clustering algorithms, we selected the aligned cluster for each utterance and used the cluster ID as a PL. With the clustering-based PLs, we can train the speaker embedding network via softmax-based objectives, analogous to supervised learning. For all our clustering benchmarks, we set the number of clusters to 5,000, which Kang et al. (2022) found to lead to the best results. For self-organizing maps, the number of clusters was set to the size of the map (7171 = 5041).

10.5.3 Input Features and Datasets

As input to all of our clustering algorithms, we employ 400-dimensional i-vectors. These compact i-vectors, which are unsupervised speaker representations, allow us to perform clustering more efficiently and avoid the high dimensionality of the MFCC acoustic features.

To evaluate the performance of our proposed clustering approach and the generated PLs for self-supervised SV, we conducted a set of experiments based on the VoxCeleb2 dataset (Chung et al. 2018). We used the development subset of the VoxCeleb2 dataset, consisting of 1,092,009 utterances collected from 5,994 speakers, to train the embedding networks. The evaluation was performed according to the original VoxCeleb1 trials list (Nagrani et al. 2017), which consists of 37,720 trials from 4,874 utterances spoken by 40 speakers.

For our ECAPA-TDNN-based SV system, we used 40-dimensional Mel-frequency cepstral coefficients (MFCCs) extracted every 10 ms with a 25 ms Hamming window via the Kaldi toolkit (Povey et al. 2011). Additionally, to align with other SV works, we used waveform-level data augmentations, including additive noise and room impulse response (RIR) simulation (Snyder et al. 2018), in training the ECAPA-TDNN-based systems. In addition to the waveform-level augmentations, we applied augmentation over the extracted MFCC features, analogous to the specaugment scheme (Park et al. 2019).

10.5.4 Clustering Models and Training Details

To improve generalization, we use the AAMSoftmax objective (Deng et al. 2018) to train our self-supervised speaker embedding network, with a scale factor $s = 30$ and an angular margin $m = 0.1$. Cosine similarity is used as a backend for verification scoring between enrollment and test embeddings.

Following the IMSAT setup, we use the same MLP-based d-S-S-C architecture, where $d = 400$ and C are the input and output dimensionalities, respectively. $S = 20, 800$ neurons represent the width of the network. We use ReLU for all the hidden activations, apply batch normalization to the hidden layers, and use Softmax in the output layer. For optimization, we use the momentum algorithm with an initial learning rate of 0.01, a momentum of 0.9, and an exponential rate decay of 0.996. We set $\lambda = 0.5$ and $\mu = 3.5$.

We use a batch size of 10,240 i-vectors, and inputs are normalized independently along the samples axis to unit l2-norm to avoid losing speaker information. We use $\alpha = 1$ as the coefficient of the Beta distribution for mixup interpolation. We ran experiments for 150 epochs using 64 CPU cores for each clustering algorithm. Additionally, all SV experiments were run for 7 days using a single RTX2080Ti GPU, with a batch size of 200 MFCC samples. All code and methods used in our experiments are based on TensorFlow.

10.5.5 Self-Supervised Additive Angular Margin Softmax Objective

The AAMSoftmax objective is one of the most popular methods for training a speaker embedding network (Deng et al. 2018), and is formulated as in Eq. (10.6):

$$\mathcal{L}_{AAMSoftmax} = -\frac{1}{N}\sum_{i=1}^{N} log(\frac{e^{s(cos(\theta_{y_i}+m))}}{K_1}), \tag{10.6}$$

where $K_1 = e^{s(cos(\theta_{y_i}+m))} + \sum_{j=1,j\neq i}^{C} e^{scos\theta_j}$, N is the batch size, C is the number of classes, y_i corresponds to label index, θ_{y_i} represents the angle between the column vector of weight matrix $W_{y_i} \in \mathbb{R}^{512\times 1}$ of the ground truth center and the ith feature embedding $x_i \in \mathbb{R}^{512\times 1}$, where both W_{y_i} and x_i are l_2 normalized. $\theta_j = \arccos({W_j}^T x_i)$ is the angle between x_i and the jth class center $W_j \in \mathbb{R}^{512\times 1}$. The scale factor s ensures that the gradient is not too small during training, and m is a hyperparameter that encourages the similarity of correct classes to be greater than that of incorrect classes by a margin m.

The training of AAMSoftmax for self-supervised speaker embedding learning is made possible by using our generated PLs, as the above objective requires speaker labels for training.

10.5.6 ECAPA-TDNN Architecture Details

Table 10.1 describes the architecture of our ECAPA-TDNN model used for SV.

10.6 Clustering Evaluation Metrics

Using commonly accepted evaluation metrics for clustering, we thoroughly assess the quality of the generated PLs from different perspectives.

We employ seven supervised metrics based on both the PLs and true labels: unsupervised clustering accuracy, normalized mutual information (Estévez et al. 2009), adjusted mutual information (Vinh et al. 2010), completeness score (Rosenberg and Hirschberg 2007), homogeneity score (Rosenberg and Hirschberg 2007), purity score, and Fowlkes-Mallows index (Fowlkes and Mallows 1983). These metrics

Table 10.1 Standard ECAPA-TDNN architecture. T indicates the duration of features in number of frames and d the feature vector dimensionality and C and N_c represent the number of channels and classes, respectively. FC and BN stand for fully connected layer and batch normalization, respectively. In this work, we use $C = 512$

Layer	Input dimension	Output dimension
Conv1d+ReLU+BN	$d \times T$	$C \times T$
SE-Res2Block	$C \times T$	$C \times T$
SE-Res2Block	$C \times T$	$C \times T$
SE-Res2Block	$C \times T$	$C \times T$
Conv1d+ReLU	$3C \times T$	$1536 \times T$
Attentive Statistics Pooling + BN	$1536 \times T$	3072×1
FC + BN	3072×1	192×1
AAMSoftmax	192×1	$N_c \times 1$

assess the following criteria: clustering accuracy and mutual information as measures of consistency between the true labels and the generated PLs; homogeneity, completeness, and purity of clusters; and precision and recall. Additionally, we compute three unsupervised metrics—Silhouette score (Rousseeuw 1987), Calinski-Harabasz score (Caliński et al. 1974), and Davies-Bouldin score (Davies and Bouldin 1979)—that are solely based on the generated PLs and the data samples. These metrics measure how compact or scattered the clusters are (e.g., intra-class dispersion, between-cluster distances, nearest-cluster distance).

More details and discussion about the clustering metrics are available in Fathan et al. (2022), which found a very high correlation between these metrics and SV performance. Additionally, using the three unsupervised clustering metrics allows us to objectively assess our clustering performance and avoid arbitrary techniques such as t-SNE visualizations (Van der Maaten and Hinton 2008). To compute these metrics, we use available implementations from the scikit-learn toolkit. Details of the clustering metrics are as follows:

Unsupervised Clustering Accuracy (ACC). Measures the consistency between the true labels and the generated PLs. $ACC = \underset{m}{max} \frac{\sum_{i=1}^{N} \mathbb{1}\{y_i = m(c_i)\}}{N}$ where y_i is the true label, c_i is the generated PL assignment, and m is a mapping function which ranges over all possible one-to-one mappings between true labels and assignments. The optimal mapping can be efficiently computed using the Hungarian algorithm (Kuhn 2005).

Normalized Mutual Information (NMI). $NMI(Y, C) = \frac{I(Y,C)}{\frac{1}{2}[H(Y)+H(C)]}$ where Y and C denote the ground-truth labels and the generated clustering assignments, respectively(Estévez et al. 2009). H is the entropy function and I denotes the MI metric. NMI is the harmonic mean between below homogeneity and completeness scores.

Adjusted MI (AMI). Adjusted MI (Vinh et al. 2010) is NMI adjusted for chance by discounting a chance normalization term. Since the NMI measure is not adjusted for chance, including the adjusted MI score might be preferred for comparison in some of our cases.

Completeness Score. A clustering assignment satisfies completeness if all the data points that are members of a given class are elements of the same cluster (Rosenberg and Hirschberg 2007). The scores are between 0 and 1, where 1 stands for perfectly complete assignment.

Homogeneity Score. A clustering assignment satisfies homogeneity if all of its clusters contain only data points that are members of a single class (Rosenberg and Hirschberg 2007). The score is between 0 and 1, where 1 stands for perfectly homogeneous assignment.

Purity Score. Each cluster is assigned to the class that is most frequent in the cluster, and then the accuracy of this assignment is measured by counting the number of correctly assigned samples and dividing by number of samples N. Cluster purity measures how pure clusters are. If a cluster is composed of members of the same class, then it is completely pure.

Fowlkes-Mallows Index (FMI). Measures the similarity of two clustering methods by computing the geometric mean between the precision and recall (Fowlkes and Mallows 1983). A higher score indicates a good similarity between two clusters.

Silhouette Score. The Silhouette score is calculated using (a) the mean intra-cluster distance and (b) the mean nearest-cluster distance for each sample (Rousseeuw 1987). The Silhouette Coefficient for a sample is $\frac{(b-a)}{\max(a,b)}$.

Davies-Bouldin Score (DBS). The average similarity measure of each cluster with its most similar cluster, where similarity is the ratio of within-cluster distances to between-cluster distances (Davies and Bouldin 1979). Thus, clusters that are farther apart and less dispersed will result in a better score. Lower values indicate better clustering.

10.7 Details of Proposed Augmentation-Based SSL Objectives

The following list provides detailed descriptions of the augmentation-based SSL objectives used in our experiments:

Data Augmentation Loss $\mathcal{L}_{aug}$. $\mathcal{L}_{aug}$ forces the predicted representations of augmented samples to be close to those of the original data points by minimizing the KL-divergence between both predictions, as in Eq. (10.7):

$$\mathcal{L}_{aug} = \frac{1}{N} \sum_{i=1}^{N} KL(p_i^{aug_{r_i}} || p_i) \tag{10.7}$$

with $J = \{aug_1, ..., aug_{|J|}\}$ the ensemble of available data augmentations and $r_i \in \{1, .., |J|\}$ that refers to a random augmentation from J. $KL(.||.)$ refers to the Kullback-Leibler divergence, and N is the size of data (or mini-batches). $p_i = f(x_i) \in \mathbb{R}^{1xC}$, and $p_i^{aug_j} = f(x_i^{aug_j})$ correspond to the predictions of data sample x_i and its augmented version $x_i^{aug_j}$, respectively.

Contrastive Self-Supervised Learning (InfoNCE). InfoNCE (Oord et al. 2018), where NCE stands for Noise-Contrastive Estimation, is a type of contrastive loss function used for self-supervised learning in SimCLR (Chen et al. 2020). It is also known as the NT-Xent loss (Normalized Temperature-scaled Cross-Entropy). The goal is to maximize the similarity between the representations of two augmented versions of the same input, Z_i and Z_j, while minimizing their similarity to all other examples in the batch. In short, the InfoNCE loss compares the similarity of Z_i and Z_j to the similarity of Z_i to any other representation in the batch by performing a Softmax over the similarity values. The InfoNCE loss $\mathcal{L}_{i,j}$ for pair (i,j) can be written as in (10.8):

$$\mathcal{L}_{i,j} = -log \frac{\exp sim(Z_i, Z_j)/\tau}{\sum_{k=1}^{2N} \mathbb{1}_{k \neq i} \exp sim(Z_i, Z_k)/\tau} \tag{10.8}$$

$\mathbb{1}_{k \neq i} \in \{0, 1\}$ is an indicator function evaluating to 1, if and only if, $k \neq i$, and $\tau = 1$ denotes the temperature parameter. The final $\mathcal{L}_{InfoNCE}$ loss is computed across all positive pairs, both (i, j) and (j, i), in a mini-batch (a sample and its augmented version). The default similarity metric that is used is cosine similarity, defined as: $sim(Z_i, Z_j) = \frac{Z_i^T . Z_j}{\|Z_i\| \|Z_j\|}$. We also studied KL-divergence as a distance metric but the results were worse than cosine distance.

Supervised Contrastive Loss (SupCon). $\mathcal{L}_{SupCon}$ from Khosla et al. (2020) extends the self-supervised batch contrastive approach of the NT-Xent loss (Normalized Temperature-scaled Cross Entropy) (Chen et al. 2020) to the fully-supervised setting, allowing us to effectively leverage label information. For that, clusters of points belonging to the same class are pulled together in normalized embedding space, while simultaneously pushing apart clusters of samples from different classes. The SupCon extension allows for multiple positives per anchor instead of a single sample in addition to many negatives, and draws from samples of the same class as the anchor, rather than being data augmentations of the anchor, as done in previous works. It showed benefits for robustness to natural corruptions and is more stable to hyperparameter settings such as optimizers and data augmentations.

Since the SupCon loss[2] requires labels, the novelty of our usage is to use online generated labels as input labels to the SupCon loss function, which allows us to use it in a completely unsupervised fashion without the need for ground-truth labels. Additionally, our framework does not rely on any additional modules such as a projection network or a separate encoder.

Variance-Invariance-Covariance Regularization (VICReg). VICReg (Bardes et al. 2021) aims to maximize the agreement between representations of augmented views of the same instance while preventing the collapse problem. It uses two regularization terms: (1) a term $\mathcal{L}_{VICReg'variance}$ that maintains the variance of each embedding dimension above a threshold; and (2) a term $\mathcal{L}_{VICReg'covariance}$ that decorrelates each pair of variables. VICReg loss is composed of a variance, invariance and covariance loss terms that are added to each other as follows: $\mathcal{L}_{VICReg} = \lambda s(Z, Z') + \mu[v(Z) + v(Z')] + \nu[c(Z) + c(Z')]$

The variance regularization term is: $v(Z) = \frac{1}{C} \sum_{j=1}^{C} \max(0, \gamma - S(Z^j, \epsilon))$ where $S(x, \epsilon) = \sqrt{Var(x) + \epsilon}$ and $\gamma = 1$ is a constant target value for the standard deviation. Z^j denotes the vector composed of all values at dimension j in all logit vectors in batch matrix Z and $\epsilon = 0.0001$ is a small scaler for stability. The covariance regularization term is computed as follows: $c(Z) = \frac{1}{C} \sum_{i \neq j} [C(Z)]_{i,j}^2$, with $C(Z) = \frac{1}{N-1} \sum_{i=1}^{N} (Z_i - \overline{Z})(Z_i - \overline{Z})^T$ and $\overline{Z} = \frac{1}{N} \sum_{i=1}^{N} Z_i$. Z_i denotes the logits of sample i. The invariance criterion is simply: $s(Z, Z') = \frac{1}{N} \sum_i \|Z_i - Z_i'\|_2^2$.

In our implementation, we use $\lambda = 1$, $\mu = 1$, and $\nu = 0.5$. Additionally, since the covariance matrix requires large memory usage, we set the batch size to 640 samples and use $C = 5000$ clusters.

[2] We use the implementation from https://github.com/wangz10/contrastive_loss/blob/master/losses.py with temperature=1 and base_temperature=1.

10.8 Results and Discussion

10.8.1 Ablation Study

In Table 10.2, we performed a large-scale ablation study to analyze the contribution of all components of our system and the influence of the predefined number of clusters. We also study the VICReg method (Bardes et al. 2021) which comprises a term $\mathcal{L}_{VICReg'variance}$ that maintains the variance of each embedding dimension above a threshold and a term $\mathcal{L}_{VICReg'covariance}$ that decorrelates each pair of variables (see Sect. 10.7 for mathematical details). Results show that there is complementary information between all loss terms in our proposed CIMC objective and that each helps boost the performance of the overall clustering framework. We also observe that choosing a much higher number of clusters than the ground-truth leads to improved clustering performance across all studied systems.

10.8.2 Comparison of CIMC to Other Clustering Benchmarks

In Table 10.3, we provide the results for a large variety of clustering benchmarks from Table 1 of Fathan et al. (2022), compared to our proposed method with and without explicit data augmentation. According to the results, our approach outperforms all other baselines in terms of clustering metrics achieving 63.9% unsupervised accuracy, while having a compute time comparable to classical clustering models (3-4 days). This is compared to 60.2% for AHC which was the best-performing method (6.2% relative improvement).

Using our proposed system's generated PLs to train our speaker embedding system allows us to achieve a very competitive downstream SV EER performance outperforming all other benchmarks, except the AHC PLs which lead to a slightly better performance. Moreover, using audio data augmentation by additionally incorporating $\mathcal{L}_{aug}$ to push two augmented versions of the same sample closer (system denoted as CIMC+ in Eq. (10.9)) and an additional $\mathcal{L}_{InfoNCE}$ loss term for contrastive learning (system denoted as CIMC++ in Eq. (10.10)) helps further enhance both clustering and downstream SV performance higher than all our studied baselines. Our CIMC++ clustering system achieved 72.5% unsupervised accuracy and 20.4% relative clustering improvement, and our combinations show that there is complementarity between our studied objectives. CIMC+ and CIMC++ clustering objectives are defined as in (10.9) and (10.10):

$$CIMC+ = \mathcal{L}_{IMSAT} + \mathcal{L}_{Mixup} + \mathcal{L}_{SupCon} + \mathcal{L}_{aug} \tag{10.9}$$

$$CIMC++ = \mathcal{L}_{IMSAT} + \mathcal{L}_{Mixup} + \mathcal{L}_{SupCon} + \mathcal{L}_{aug} + \mathcal{L}_{InfoNCE} \tag{10.10}$$

Table 10.2 An ablation study of our proposed CIMC clustering system including various SSL-based loss objectives that do not employ data augmentation (only original data samples). C denotes the predefined number of clusters. Results are reported in terms of clustering metrics and the corresponding Equal Error Rate (EER) downstream SV evaluation performance when using the generated PLs to train our studied SV system

Model	Clustering metrics											Speaker verification
	ACC	AMI	NMI	No. of clusters	Completeness	Homogeneity	FMI	Purity	Silhouette	CHS	DBS	EER (%)
$\mathcal{L}_{Mixup}$ (C: 10k)	0.013	0.016	0.432	10000	0.413	0.452	0.001	0.026	–0.019	1.001	17.633	9.767
$\mathcal{L}_{SupCon}$ (C: 10k)	0.015	0.02	0.419	10000	0.404	0.434	0.001	0.027	–0.036	1.001	19.5	20.074
$\mathcal{L}_{VICReg}$ (Bardes et al. 2021) (C: 5k)	0.018	0.082	0.27	4496	0.309	0.239	0.004	0.021	–0.134	1.001	18.031	11.612
$\mathcal{L}_{IMSAT}$ (C: 5k)	0.578	0.731	0.822	5000	0.83	0.815	0.552	0.604	–0.033	**1.002**	26.56	4.507
$\mathcal{L}_{IMSAT}$ (C: 5994)	0.600	0.743	0.833	5993	0.834	0.831	0.583	0.636	–0.074	0.999	23.915	4.295
$\mathcal{L}_{IMSAT}$ (C: 10k)	0.621	0.754	0.844	9844	0.836	0.853	0.616	0.678	–0.122	0.999	16.897	4.438
$\mathcal{L}_{Mixup} + \mathcal{L}_{SupCon}$ (C: 10k)	0.015	0.034	0.354	9639	0.36	0.348	0.002	0.023	–0.133	0.999	**15.563**	12.54
$\mathcal{L}_{IMSAT} + \mathcal{L}_{Mixup} + \mathcal{L}_{VICReg'variance} + \mathcal{L}_{VICReg'covariance}$ (C: 5k)	0.013	0.018	0.369	5000	0.367	0.371	0.001	0.017	**–0.015**	1.0	25.571	19.952
$\mathcal{L}_{IMSAT} + \mathcal{L}_{Mixup} + \mathcal{L}_{SupCon} + \mathcal{L}_{VICReg'variance} + \mathcal{L}_{VICReg'covariance}$ (C: 5k)	0.014	0.02	0.36	5000	0.361	0.359	0.001	0.017	–0.022	0.999	26.667	21.84
$\mathcal{L}_{IMSAT} + \mathcal{L}_{Mixup}$ (C: 10k)	0.628	0.764	0.852	9791	0.841	0.862	0.615	0.692	–0.149	1.0	17.297	4.321

(continued)

Table 10.2 (continued)

Model	Clustering metrics											Speaker verification
	ACC	AMI	NMI	No. of clusters	Completeness	Homogeneity	FMI	Purity	Silhouette	CHS	DBS	EER (%)
$\mathcal{L}_{IMSAT} + \mathcal{L}_{SupCon}$ (C: 5k)	0.497	0.688	0.784	4996	0.81	0.76	0.347	0.516	–0.065	0.999	24.809	4.623
$\mathcal{L}_{IMSAT} + \mathcal{L}_{SupCon}$ (C: 5994)	0.514	0.697	0.793	5974	0.814	0.774	0.347	0.538	–0.117	0.999	22.164	4.475
$\mathcal{L}_{IMSAT} + \mathcal{L}_{SupCon}$ (C: 10k)	0.548	0.717	0.813	9585	0.823	0.803	0.361	0.589	–0.138	1.001	15.941	4.348
$CIMC = \mathcal{L}_{IMSAT} + \mathcal{L}_{Mixup} + \mathcal{L}_{SupCon}$ (C: 5k)	0.602	0.751	0.836	4999	0.842	0.831	0.579	0.632	–0.071	0.999	26.905	**4.231**
$CIMC = \mathcal{L}_{IMSAT} + \mathcal{L}_{Mixup} + \mathcal{L}_{SupCon}$ (C: 5994)	0.619	0.761	0.846	5989	0.845	0.846	0.6	0.66	–0.125	1.002	24.301	4.321
$CIMC = \mathcal{L}_{IMSAT} + \mathcal{L}_{Mixup} + \mathcal{L}_{SupCon}$ (C: 10k)	**0.639**	**0.776**	**0.86**	9685	**0.8 47**	**0.873**	**0.642**	**0.71**	–0.136	0.998	17.599	4.252

(continued)

Table 10.2 (continued)

Model	Clustering metrics											Speaker verification
	ACC	AMI	NMI	No. of clusters	Completeness	Homogeneity	FMI	Purity	Silhouette	CHS	DBS	EER (%)
$CIMC+ = \mathcal{L}_{IMSAT} + \mathcal{L}_{Mixup} + \mathcal{L}_{SupCon} + \mathcal{L}_{aug}$ (C: 10k)	0.714	0.834	0.894	7810	0.887	0.901	0.728	0.773	**–0.129**	0.999	19.768	3.377
$CIMC{+}{+} = \mathcal{L}_{IMSAT} + \mathcal{L}_{Mixup} + \mathcal{L}_{SupCon} + \mathcal{L}_{aug} + \mathcal{L}_{InfoNCE}$ (C: 10k)	**0.725**	**0.842**	**0.9**	8500	**0.89**	**0.91**	**0.746**	**0.792**	–0.134	**1.0**	18.407	**3.362**

Table 10.3 A comparison study of our proposed clustering methods CIMC and CIMC++ compared to a large set of benchmarks (classical and deep-learning based models). Results are reported in terms of Clustering performance (clustering metrics) and the corresponding EER (%) downstream SV evaluation performance when using the generated PLs to train our studied SV system. l2-norm refers to normalizing i-vector inputs independently along the samples axis to unit l2-norm instead of mean and standard deviation scaling (StandardScaler) of i-vectors along the features axis. Both CIMC and CIMC++ employ l2-norm

Model	Clustering metrics											Speaker verification
	ACC	AMI	NMI	No. of clusters	Completeness	Homogeneity	FMI	Purity	Silhouette	CHS	DBS	EER (%)
Supervised (True Labels)	**1.0**	**1.0**	**1.0**	5994	**1.0**	**1.0**	**1.0**	**1.0**	**–0.006**	31.708	4.692	**1.437**
GMM (Full cov.)	0.45	0.631	0.747	5000	0.767	0.728	0.312	0.566	–0.015	39.266	4.673	5.143
GMM (Full cov., l2-norm)	0.504	0.678	0.789	5000	0.792	0.785	0.415	0.633	–0.015	41.568	5.114	5.429
Bayesian GMM (γ=1e-5, μ=1)	0.45	0.629	0.746	5000	0.766	0.727	0.312	0.566	–0.015	39.257	4.673	5.143
Bayesian GMM (l2-norm, γ=1e-5, μ=1)	0.504	0.678	0.789	5000	0.792	0.785	0.415	0.633	–0.015	**41.57**	5.115	5.159
Divisive HC	0.097	0.204	0.477	5000	0.479	0.474	0.035	0.132	–0.06	18.044	9.068	13.531
KMeans	0.302	0.468	0.591	5000	0.645	0.546	0.194	0.311	–0.114	24.936	**2.714**	6.978
CURE	0.151	0.218	0.393	5000	0.466	0.34	0.011	0.216	–0.052	17.77	5.372	6.994
BIRCH	0.299	0.374	0.54	5000	0.725	0.43	0.013	0.353	–0.027	24.348	4.901	5.642
DEC	0.029	0.122	0.365	4911	0.386	0.345	0.007	0.036	–0.084	8.734	7.266	11.957
SOM	0.025	0.088	0.402	5041	0.404	0.4	0.01	0.037	–0.041	10.148	18.402	15.806
DeepCWRN	0.003	0.006	0.15	1008	0.179	0.129	0.001	0.003	–0.217	3.841	41.521	38.171
IMSAT	0.393	0.491	0.649	4987	0.668	0.63	0.297	0.426	–0.044	22.887	6.668	5.912

(continued)

Table 10.3 (continued)

Model	Clustering metrics											Speaker verification
	ACC	AMI	NMI	No. of clusters	Completeness	Homogeneity	FMI	Purity	Silhouette	CHS	DBS	EER (%)
AHC	0.587	0.74	0.825	5000	0.841	0.81	0.311	0.684	**–0.01**	39.561	4.991	3.685
AHC (l2-norm)	0.602	0.756	0.838	5000	0.849	0.827	0.375	0.693	–0.034	39.638	5.147	3.621
CIMC (w/o data augmentation)	0.639	0.776	0.86	9685	0.847	0.873	0.642	0.71	–0.136	0.998	17.599	4.252
CIMC++ (w/ data augmentation)	**0.725**	**0.842**	**0.9**	8500	**0.89**	**0.91**	**0.746**	**0.792**	–0.134	1.0	18.407	**3.362**

Definitions and details of all these loss objectives are included in Sect. 10.7.

10.8.3 The Evolution of Clustering Metrics Over Time

In Fig. 10.3, we show the evolution of clustering metrics and the number of clusters discovered during the training process. Results show that regularization through data augmentation helps improve performance considerably, and that using augmentations through objectives $\mathcal{L}_{aug}$ or $\mathcal{L}_{InfoNCE}$ takes more epochs to achieve the best clustering performance. As incorporating these objectives also consumes more computing resources, this results in the whole training process taking around 10 times longer for training compared to our proposed augmentation-free clustering approach which only requires around 3 days to converge to its best performance.

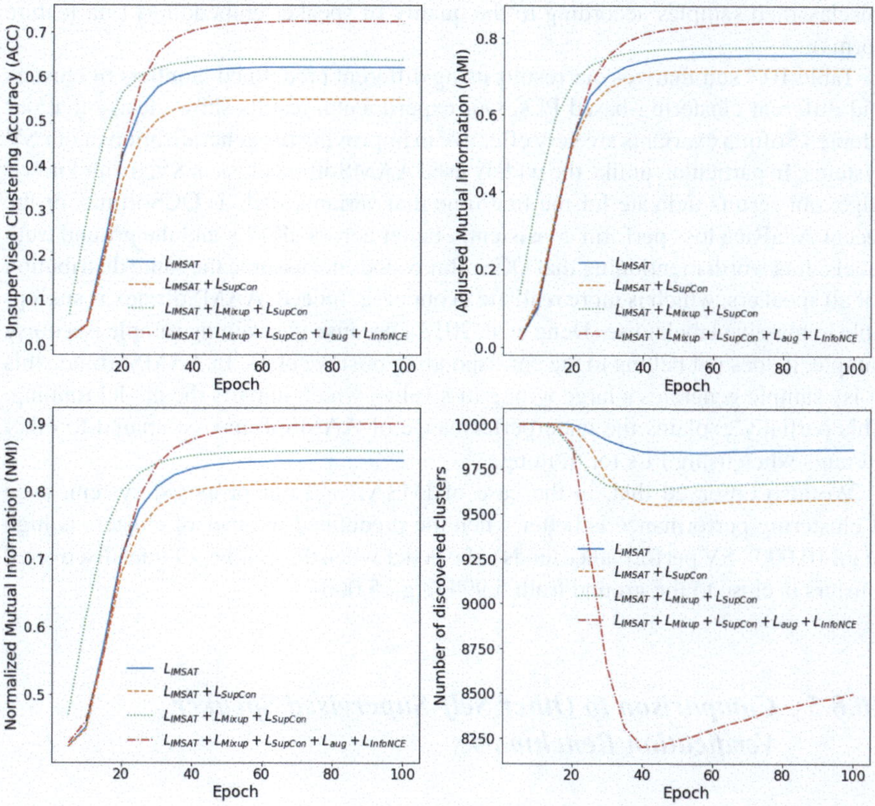

Fig. 10.3 The evolution of clustering metrics over epochs and the number of clusters discovered during training of our clustering systems based on various loss combinations

10.8.4 Study of Various Maximum Margin-Based Objectives

To provide a clear geometric interpretation of data samples and enhance the discriminative power of deep models, AAMSoftmax objective (Deng et al. 2018) (also known as ArcFace) introduces an additive angular margin to the target angle (between the given features and the target center). Due to the exact correspondence between the angle and arc in the normalized hypersphere, AAMSoftmax can directly optimize the geodesic distance margin, which is why it is also called ArcFace.

Additionally, CosFace (Wang et al. 2018b), a large-margin cosine loss, reformulates the Softmax loss as a cosine loss by L2 normalizing both features and weight vectors to remove radial variations, based on which a cosine margin term is introduced to further maximize the decision margin in the angular space. In contrast, OCSoftmax (Zhang et al. 2021) uses one-class learning instead of multi-class classification and does not assume the same distribution for all classes/speakers. More recently, AdaFace loss (Kim et al. 2022) has been proposed which emphasizes misclassified samples according to the quality of speaker embeddings (via feature norms).

Table 10.4 summarizes our results using different predefined numbers of clusters and different clustering-based PLs. Our experimental results show clearly that our adopted Softmax variants are very effective in improving the generalization of our SV systems. In particular, unlike the widely used AAMSoftmax loss in SV, to our knowledge, our results indicate for the first time that variants such as OCSoftmax or the recent AdaFace loss perform consistently better across all PLs and the ground truth labels. It is worth mentioning that OCSoftmax does not assume the same distribution for all speakers, which is more realistic in our case. Indeed, AAMSoftmax is susceptible to massive label noise (Deng et al. 2018), because if a training sample is a noisy sample, it does not belong to the corresponding positive class. In AAMSoftmax, this noisy sample generates a large wrong loss value, which impairs the model training. This partially explains the underperformance of AAMSoftmax compared to other variants when using PLs for training.

We also observed that, in the case of IMSAT and our proposed system, even if clustering performance is better when the predefined number of clusters is high (e.g., 10,000), SV performance tends to be better when the number of final discovered clusters is close to the ground truth 5,994 (e.g., 5,000).

10.8.5 Comparison to Other Self-Supervised Speaker Verification Benchmarks

Finally, Table reftab:sslspsbaselinesspstable shows a comparison of our approach with and without Data Augmentation (DA) for Self-Supervised SV (SSSV) training using our system-based PLs compared to recent SOTA SSSV approaches employing diverse SSL objectives with the same ECAPA-TDNN model encoder. The results

Table 10.4 A study of various margin-based Softmax losses for better generalization of our SV system, using different pseudo-labels. AAMSoftmax is by Deng et al. (2018), AMSoftmax by Wang et al. (2018a), OCSoftmax by Zhang et al. (2021), AdaFace by Kim et al. (2022), and CosFace by Wang et al. (2018b)

Pseudo-labels	No. clusters	AAMSoftmax	AMSoftmax	OCSoftmax	AdaFace	CosFace	Cross entropy
True labels	5,994	1.437	1.522	1.416	**1.326**	1.463	3.489
GMM	5,000	5.429	4.851	**4.682**	5.095	4.862	8.425
AHC	5,000	3.621	3.664	3.584	**3.526**	3.6	6.479
IMSAT	5,000	4.507	4.141	3.881	**3.807**	4.083	7.206
	5,994	4.146	3.961	3.892	4.008	**3.696**	7.333
	10,000	4.438	4.024	**4.003**	4.046	4.072	7.370
$\mathcal{L}_{IMSAT} + \mathcal{L}_{SupCon}$	5,000	4.623	4.401	**4.396**	4.576	4.48	7.179
	5,994	4.475	4.502	4.491	4.443	**4.427**	7.179
	10,000	4.348	4.221	4.189	4.343	**4.173**	7.174
CIMC =	5,000	4.231	4.046	**3.924**	4.056	4.332	7.391
$\mathcal{L}_{IMSAT} + \mathcal{L}_{Mixup}$	5,994	4.321	4.146	**4.024**	4.125	4.199	7.28
$+\mathcal{L}_{SupCon}$	10,000	4.252	**4.03**	4.146	4.21	4.093	7.259
CIMC+ = CIMC + $\mathcal{L}_{aug}$	10,000	3.377	**3.287**	3.293	3.399	3.298	5.695
CIMC++ = CIMC + $\mathcal{L}_{InfoNCE}$	10,000	3.362	**3.001**	3.006	3.043	3.181	5.801

Table 10.5 Some recent SOTA self-supervised SV approaches in EER (%) compared to our simple SV system trained with our PLs. All models are based on ECAPA-TDNN. Results are reported on the original VoxCeleb1 test set (Voxceleb1_O)

SSL objective	EER (%)
MoBY (Xia et al. 2021)	8.2
InfoNCE (Tao et al. 2022)	7.36
MoCo (Cho et al. 2021)	7.3
ProtoNCE (Xia et al. 2021)	7.21
PCL (Xia et al. 2021)	7.11
CA-DINO (Han et al. 2023)	3.585
i-mix (Fathan and Alam 2023)	3.478
l-mix (Fathan and Alam 2023)	3.377
Iterative clustering (Tao et al. 2022)	3.09
Ours w/o DA (CIMC & OCSoftmax)	**3.924**
Ours w/ DA (CIMC++ & AMSoftmax)	**3.001**

show clearly that our CIMC approach provides very competitive performance while being simple and fast. Besides, when employing augmentations in CIMC++, our approach outperforms all the baselines, which suggests that regularization through DA is still crucial and that further gains can be made by simply improving the clustering modules of current self-supervised speaker recognition systems (Table 10.5).

10.9 Conclusion

In this chapter, we proposed a general-purpose and multi-objective clustering method. Our approach avoids using explicit data augmentation for fast and efficient training. It is based on three principles: (1) self-augmented training to enforce representation invariance and maximize the information-theoretic dependency between samples and their predicted pseudo-labels; (2) virtual mixup training to impose local-Lipschitzness which enforces the cluster assumption; and (3) supervised contrastive learning by leveraging on-the-fly generated pseudo-labels to pull samples of same class together and push samples of different clusters apart. Moreover, we explored various recent SOTA SSL objectives for clustering, including two data augmentation-based objectives, where we showed that our multi-objective approach provides beneficial complementary information. Our approach outperformed all other baselines used to cluster speaker embeddings and provided very competitive SV performance outperforming all the benchmarks.

Acknowledgements The authors wish to acknowledge the funding from the Government of Canada's New Frontiers in Research Fund (NFRF) through grant NFRFR-2021-00338 and Natural Sciences and Engineering Research Council of Canada (NSERC) through grant RGPIN-2019-05381. Any opinions, findings, conclusions or recommendations expressed in this material are those of the authors and do not necessarily reflect those of the NFRF and NSERC.

References

Arpit D et al (2017) A closer look at memorization in deep networks. In: ICML

Bahri D et al (2021) Scarf: self-supervised contrastive learning using random feature corruption. arXiv preprint arXiv:2106.15147

Bai Z, Zhang XL (2021) Speaker recognition based on deep learning: an overview. Neural Netw

Bardes A, Ponce J, LeCun Y (2021) Vicreg: variance-invariance-covariance regularization for self-supervised learning. arXiv preprint arXiv:2105.04906

Bridle J et al (1991) Unsupervised classifiers, mutual information and'phantom targets. In: Conference on neural information processing systems (NeurIPS) 4

Caliński T et al (1974) A dendrite method for cluster analysis. Commun Stat-Theory Methods

Chen T, Kornblith S, Norouzi M, Hinton G (2020) A simple framework for contrastive learning of visual representations. In: Proceedings of the 37th international conference on machine learning, PMLR, Proceedings of Machine Learning Research, vol 119, pp 1597–1607. https://proceedings.mlr.press/v119/chen20j.html

Cho J et al (2021) The jhu submission to voxsrc-21: track 3. arXiv preprint arXiv:2109.13425

Chung JS, Nagrani A, Zisserman A (2018) Voxceleb2: deep speaker recognition. In: The International speech communication association (INTERSPEECH)

Dahal P (2018) Learning embedding space for clustering from deep representations. In: IEEE big data

Davies DL, Bouldin DW (1979) A cluster separation measure. IEEE Trans PAMI

Day WH et al (1984) Efficient algorithms for agglomerative hierarchical clustering methods. J Classif

Dehak N et al (2011) Front-end factor analysis for speaker verification. IEEE/ACM Trans Audio Speech Lang

Deng J, Guo J, Zafeiriou S (2018) Arcface: additive angular margin loss for deep face recognition. In: 2019 IEEE/CVF conference on computer vision and pattern recognition (CVPR), pp 4685–4694. https://api.semanticscholar.org/CorpusID:8923541

Desplanques B et al (2020) ECAPA-TDNN: emphasized channel attention, propagation and aggregation in TDNN based speaker verification. In: The international speech communication association (INTERSPEECH). ISCA

Dilokthanakul N et al (2016) Deep unsupervised clustering with gaussian mixture variational autoencoders. arXiv preprint arXiv:1611.02648

Dosovitskiy A et al (2014) Discriminative unsupervised feature learning with convolutional neural networks. In: Conference on neural information processing systems (NeurIPS)

Estévez PA et al (2009) Normalized mutual information feature selection. IEEE Trans Neural Netw

Fathan A, Alam J (2023) On the influence of the quality of pseudo-labels on the self-supervised speaker verification task: a thorough analysis. In: IWBF. IEEE

Fathan A, Alam J, Kang W (2022) On the impact of the quality of pseudo-labels on the self-supervised speaker verification task. In: NeurIPS ENLSP Workshop

Fowlkes EB, Mallows CL (1983) A method for comparing two hierarchical clusterings. J Am Stat Assoc 78(383):553–569

Grandvalet Y, Bengio Y (2004) Semi-supervised learning by entropy minimization. In: Conference on neural information processing systems (NeurIPS) 17

Guha S et al (1998) Cure: an efficient clustering algorithm for large databases. SIGMOD Rec. https://doi.org/10.1145/276305.276312

Han B, Chen Z, Qian Y (2022) Self-supervised speaker verification using dynamic loss-gate and label correction. arXiv preprint arXiv:2208.01928

Han B et al (2023) Self-supervised learning with cluster-aware-dino for high-performance robust speaker verification. arXiv preprint arXiv:2304.05754

Hansen JH, Hasan T (2015) Speaker recognition by machines and humans: a tutorial review. IEEE Signal Process Mag 32(6):74–99

Hartigan JA, Wong MA (1979) A k-means clustering algorithm. In: Applied statistics, JSTOR, pp 100–108
Hu W, Miyato T, Tokui S, Matsumoto E, Sugiyama M (2017) Learning discrete representations via information maximizing self-augmented training. In: International conference on machine learning (ICML), PMLR, pp 1558–1567
Jiang Z et al (2016) Variational deep embedding: a generative approach to clustering. CoRR, arXiv: abs/1611.05148
Kang WH, Alam J, Fathan A (2022) l-mix: a latent-level instance mixup regularization for robust self-supervised speaker representation learning. JSTSP
Kenny P (2012) A Small Footprint I-vector Extractor. In: Odyssey, pp 1–6
Khosla P, Teterwak P et al (2020) Supervised contrastive learning. In: Conference on neural information processing systems (NeurIPS)
Kim M, Jain AK, Liu X (2022) Adaface: quality adaptive margin for face recognition. In: Proceedings of the IEEE/CVF conference on computer vision and pattern recognition, pp 18750–18759
Kuhn HW (2005) The hungarian method for the assignment problem. Naval Res Logist (NRL) 52(1):7–21
Li Y et al (2021) Contrastive clustering. In: AAAI
Van der Maaten L, Hinton G (2008) Visualizing data using t-sne. J Mach Learn Res 9(11)
Mao X et al (2019) Virtual mixup training for unsupervised domain adaptation. arXiv preprint arXiv:1905.04215
Meng L et al (2021) Mixspeech: data augmentation for low-resource automatic speech recognition. In: International conference on acoustics, speech and signal processing (ICASSP)
Miyato T et al (2018) Virtual adversarial training: a regularization method for supervised and semi-supervised learning. PAMI
Nagrani A et al (2017) Voxceleb: a large-scale speaker identification dataset. In: The international speech communication association (INTERSPEECH)
Oord Avd, Li Y, Vinyals O (2018) Representation learning with contrastive predictive coding. arXiv preprint arXiv:1807.03748
Park DS et al (2019) Specaugment: a simple data augmentation method for automatic speech recognition. In: The international speech communication association (INTERSPEECH)
Peng J et al (2022) Progressive contrastive learning for self-supervised text-independent speaker verification. In: Proceedings of Odyssey workshop
Povey D, Ghoshal A, Boulianne G, Burget L, Glembek O, Goel NK, Hannemann M, Motlícek P, Qian Y, Schwarz P, Silovský J, Stemmer G, Veselý K (2011) The kaldi speech recognition toolkit. In: Semantic scholar. https://api.semanticscholar.org/CorpusID:1774023
Ronen M et al (2022) Deepdpm: deep clustering with an unknown number of clusters. In: Proceedings of IEEE/CVF
Rosenberg A, Hirschberg J (2007) V-measure: a conditional entropy-based external cluster evaluation measure. In: Proceedings of EMNLP-CoNLL, pp 410–420
Rousseeuw PJ (1987) Silhouettes: a graphical aid to the interpretation and validation of cluster analysis. J Comput Appl Math 20:53–65
Snyder D et al (2018) X-vectors: robust dnn embeddings for speaker recognition. In: IEEE-CASSP
Tao R et al (2022) Self-supervised speaker recognition with loss-gated learning. In: International conference on acoustics, speech, and signal processing (ICASSP), IEEE
Tomilov A et al (2021) STC antispoofing systems for the ASVspoof2021 challenge. In: Proceedings of 2021 edition of the automatic speaker verification and spoofing countermeasures challenge, pp 61–67
Vapnik VN et al (2015) On the uniform convergence of relative frequencies of events to their probabilities. In: Measures of complexity. Springer, pp 11–30
Verma V et al (2022) Interpolation consistency training for semi-supervised learning. Neural Netw 145:90–106

Vinh NX, Epps J, Bailey J (2010) Information theoretic measures for clusterings comparison: variants, properties, normalization and correction for chance. J Mach Learn Res 11(95):2837–2854. http://jmlr.org/papers/v11/vinh10a.html

Wang F et al (2018) Additive margin softmax for face verification. IEEE Signal Process Lett 25(7):926–930

Wang H, Wang Y, Zhou Z, Ji X, Gong D, Zhou J, Li Z, Liu W (2018b) Cosface: large margin cosine loss for deep face recognition. In: Proceedings of the IEEE conference on computer vision and pattern recognition, pp 5265–5274

Xia W et al (2021) Self-supervised text-independent speaker verification using prototypical momentum contrastive learning. In: International conference on acoustics, speech, and signal processing (ICASSP), IEEE

Xie J et al (2016) Unsupervised deep embedding for clustering analysis. In: ICML, PMLR, pp 478–487

Zhang H et al (2017) mixup: beyond empirical risk minimization. arXiv preprint arXiv:1710.09412

Zhang T, Ramakrishnan R, Livny M (1997) BIRCH: a new data clustering algorithm and its applications. Data Min Knowl Disc 1:141–182

Zhang Y et al (2021) One-class learning towards synthetic voice spoofing detection. IEEE Signal Process Lett

Part VI
Conclusion

Chapter 11
Remaining Issues for AI

Andy Way, Peyman Passban, and Mehdi Rezagholizadeh

Abstract In this book, we have featured a number of techniques which can be hugely beneficial for AI practitioners. Given that almost everyone in the field—whether in academia or in industry—is using LLMs, we expect the impact of this collection to be significant. Up to this point, we have focused on providing an almost entirely positive view of AI. AI is on everyone's lips; it is on the front pages of our newspapers and magazines, it is on our TV channels, it is concerning governments right across the planet, and it is the topic of conversations at the dinner table. However, it is right to acknowledge a range of issues that require some pause for thought. Accordingly, in this chapter, we discuss some of the major concerns surrounding the widespread adoption of AI, as a counterpoint to the preceding chapters. It is not that we believe that AI cannot be used for good, far from it; but unless the following concerns are satisfactorily addressed, it is likely that an AI winter will set in.

11.1 Pushback Against AI

Here are several reasons why we believe a balanced approach to AI development is necessary. For more in-depth material, review (Way 2024).

Exclusivity: The importance of open-sourcing and making AI accessible to everyone cannot be overstated. Currently, large companies and a few universities dominate the development of LLMs. However, democratizing AI to be available for everyone is crucial. This book focuses on the efficiency of LLMs, which is a critical tool for making LLMs more accessible to researchers and the public.

Environmental Impacts: Massive models are responsible for significant electricity consumption and CO_2 emissions (Strubell et al. 2019; Jooste et al. 2022a, b). Although big technology companies have made strides in powering their data centers

A. Way (✉) · P. Passban
ADAPT Centre, School of Computing, Dublin City University, Dublin, Ireland
e-mail: andy.way@adaptcentre.ie

M. Rezagholizadeh
Noah's Ark Lab, Huawei, Markham, ON, Canada

© The Author(s), under exclusive license to Springer Nature Switzerland AG 2025
P. Passban et al. (eds.), *Enhancing LLM Performance*, Machine Translation: Technologies and Applications 7, https://doi.org/10.1007/978-3-031-85747-8_11

and operations with renewable, green energy, it is not enough to tackle the problem, especially given the increasing use of large and deep neural models.[1]

Social Impacts, Biases and Fairness: LLMs often suffer from biases inherited from their training data, leading to considerable discrimination across genders, religions, nationalities, cultures, and languages (Chu et al. 2024). In multilingual LLMs, there is a new bias: the dominance of English training data, which can lead to improper target-language output and cultural insensitivity (Moorkens et al. 2024; Naous et al. 2023). More of these biases are being discovered with the growth of LLMs so this is hardly likely to be the last/only one.

We need to be aware of these biases and learn how to mitigate them. Societies should manage the concerns and potential harms of AI by setting rules and regulations for developing and using AI systems. One example, is the EU AI Act. We also see other AI regulations in the US that aim to prevent or resolve issues. The ongoing court case of NY Times vs. OpenAI and Microsoft, is one of them, where the NY Times claims that OpenAI used their copyrighted content without permission to train their AI models.[2] More recently, academic authors have discovered to their alarm that the contents of their books have been sold by Taylor & Francis to Microsoft for the training of their AI models, without consultation or compensation.[3]

Ethical Issues and Privacy Concerns: LLMs should provide a safe tool for users by securing their privacy and using appropriate language in interactions. There are ongoing debates and concerns regarding the safety of these models (open-source or closed-source) that need to be addressed. Companies must reassure their customers that they do not reveal data to third-party AI companies or use private data for training AI models. For example, Adobe recently changed their terms of use, leading to significant pushback and a subsequent clarification that they have never trained generative AI on customer content or allowed access to customer content beyond legal requirements.[4] New privacy and copyright issues continue to emerge, especially in creative industries (such as music labels).[5]

Hallucination: LLMs are prone to hallucination while generating text, with some claiming that the output of LLMs is often just "bullshit" (Hicks et al. 2024. Hallucination can mislead users and, in serious cases, lead to legal, ethical, or safety problems (Bommasani et al., 2021)). It is vital to understand the underlying factors of hallucination in LLMs and be able to detect and mitigate them (Tonmoy et al. 2024). While some see hallucination as a feature that can be beneficial, others adopt these models uncritically, failing to verify the accuracy of the generated outputs and propagating misinformation. Some refer to this phenomenon as "confabulation",

[1] https://www.theguardian.com/sustainable-business/greenpeace-report-google-facebook-apple-green-data-centers.

[2] https://www.nytimes.com/2023/12/27/business/media/new-york-times-open-ai-microsoft-lawsuit.html.

[3] https://www.thebookseller.com/news/academic-authors-shocked-after-taylor--francis-sells-access-to-their-research-to-microsoft-ai.

[4] https://blog.adobe.com/en/publish/2024/06/10/updating-adobes-terms-of-use.

[5] https://www.bbc.com/news/articles/ckrrr8yelzvo.

suggesting that deeper investigation could discern the truth from the hallucinatory results (see Chapter 1, Sect. 1.3.3 for more details). Regardless, we should not have to question the validity of an LLM-generated claim multiple times to determine its truthfulness. This issue, whether seen as a bug or a feature, requires serious attention.

Reliability and Accountability: Companies like Microsoft have invested heavily in AI and will have to try to monetize their offerings. However, people might be reluctant to pay for AI services, and the reliability and consistency of some of these services are also under question.[6] For example, OpenAI stopped supporting Codex in March 2023[7] and discontinued support for some models under Azure OpenAI.[8] Even for available services, our reliance on AI systems for various tasks is directly correlated with the size and quality of their training data. For most NLP applications, massive amounts of training data are simply not available, making any claims of 'solved problems' exceedingly premature.

AI Misuse: AI is not always used for good. Examples include using AI to "resurrect" deceased musicians for duets,[9] creating deepfake content, such as fake interviews with celebrities,[10] and generating AI-created images to win photography prizes.[11] Notable figures like Geoffrey Hinton have left companies like Google, warning about the dangers of AI, highlighting the severity of these issues.[12] Additionally, the use of ChatGPT has been declining (see Way 2024, with further details available in Fishkin 2023), due to similar concerns.

We could continue discussing these issues and dissecting each of them over several pages. However, this book focuses primarily on the technical aspects of AI. Nonetheless, it is crucial to at least review some of these challenges. The technical techniques mentioned in the previous chapters can be applied to address some of these issues to some extent. For example, not all models need to be large. We advocate for 'fit-for-purpose' models, which might be small, distilled LMs, for instance. All the efficiency and effectiveness techniques suggested in the previous chapters can also help reduce the carbon footprint of LLMs and/or make them faster.

LLMs are still relatively new, and the community seems to be more focused on their capabilities rather than these other issues. However, we can already see a strong desire from the community to also address and consider non-technical aspects of these models, such as those included in this chapter. However, we strongly believe that despite this growing awareness, it is nowhere near enough, and considerably more effort is needed to tackle these challenges in a comprehensive manner.

---

[6] While working on this chapter, we found that ChatGPT was down or unresponsive to our queries (July 24, 2024, 11:42 AM, Toronto, Canada).

[7] https://news.ycombinator.com/item?id=35242069.

[8] https://learn.microsoft.com/en-us/azure/ai-services/openai/concepts/legacy-models.

[9] https://www.bbc.com/news/articles/cn00e695lpvo.

[10] https://www.theguardian.com/sport/2023/apr/22/michael-schumacher-formula-one-interview-die-aktuelle-editor-sacked.

[11] https://www.theguardian.com/technology/2023/apr/17/photographer-admits-prize-winning-image-was-ai-generated.

[12] https://www.theguardian.com/technology/2023/may/02/geoffrey-hinton-godfather-of-ai-quits-google-warns-dangers-of-machine-learning.

11.2 Conclusion

The introduction of neural techniques has led to a dramatic improvement in quality for all tasks, so AI and LLMs are here to stay for the considerable future. They have some issues, such as size, compute time, and storage, but many of the techniques included in this book can help make them more efficient. We firmly believe that AI has the potential to be transformative, but for the public to be brought along, and see AI as a power for good, the other issues raised in this chapter will need to be dealt with, as well as those yet to emerge. Whether that is forced upon the field via legislation, or by a general consensus about the way the field should move forward together, we are confident that these topics can be dealt with to the satisfaction of the wider society, so that people can see AI helping them individually in a range of areas of real importance: their health; securing their finances; creating digital twins which can deal with a lot of the problems associated with living in the 21st century, and freeing them up more generally to spend time on more worthwhile activities. This is an exciting time to be working in AI, yet despite the huge impacts delivered so far, there remain a host of issues still to work on, so that in the years to come, we can see that we were directly engaged in the most exciting developments in all of human history.

References

Bommasani R, Hudson DA, Adeli E, Altman R, Arora S, von Arx S, Bernstein MS, Bohg J, Bosselut A, Brunskill E et al (2021) On the opportunities and risks of foundation models. arXiv preprint arXiv:2108.07258

Chu Z, Wang Z, Zhang W (2024) Fairness in large language models: a taxonomic survey. arXiv preprint arXiv:2404.01349

Fishkin R (2023) We analyzed millions of ChatGPT user sessions: visits are down 29% since may, programming assistance is 30% of use. Sparktoro blog https://sparktoro.com/blog/we-analyzed-millions-of-chatgpt-user-sessions-visits-are-down-29-since-may-programming-assistance-is-30-of-use/

Hicks MT, Humphries J, Slater J (2024) ChatGPT is bullshit. Ethics Inf Technol 26(38):10

Jooste W, Haque R, Way A (2022a) Knowledge distillation: a method for making neural machine translation more efficient. Information 13(2). https://www.mdpi.com/2078-2489/13/2/88

Jooste W, Way A, Haque R, Superbo R (2022b) Knowledge distillation for sustainable neural machine translation. In: Proceedings of the 15th biennial conference of the association for machine translation in the Americas (Volume 2: Users and providers track and government track), pp 221–230

Moorkens J, Way A, Lankford S (2024) Translation and Automation. Routledge, Abingdon, Oxon., UK, forthcoming

Naous T, Ryan MJ, Xu W (2023) Having beer after prayer? measuring cultural bias in large language models. ArXiv arXiv:abs/2305.14456. https://api.semanticscholar.org/CorpusID:258865272

Strubell E, Ganesh A, McCallum A (2019) Energy and policy considerations for deep learning in NLP. In: Proceedings of the 57th annual meeting of the association for computational linguistics, ACL, Florence, Italy, pp 3645–3650

Tonmoy S, Zaman S, Jain V, Rani A, Rawte V, Chadha A, Das A (2024) A comprehensive survey of hallucination mitigation techniques in large language models. arXiv preprint arXiv:2401.01313
Way A (2024) What does the future hold for translation technologies in society? In: Baumgarten S, Tieber M (eds) Routledge handbook of translation technology and society. Routledge, Abingdon. Oxon, UK

The manufacturer's authorised representative in the EU is Springer Nature Customer Service Centre GmbH, Europaplatz 3, 69115 Heidelberg, Germany. If you have any concerns regarding our products, please contact ProductSafety@springernature.com

Printed and bound by CPI Group (UK) Ltd, Croydon, CR0 4YY
30/09/2026
02225572-0003

The manufacturer's authorised representative in the EU is Springer Nature Customer Service Centre GmbH, Europaplatz 3, 69115 Heidelberg, Germany. If you have any concerns regarding our products, please contact ProductSafety@springernature.com

Printed and bound by CPI Group (UK) Ltd, Croydon, CR0 4YY
30/09/2026
02225572-0003